UNDERSTANDING EXPERIMENTAL DESIGN AND INTERPRETATION IN PHARMACEUTICS

ELLIS HORWOOD SERIES IN PHARMACEUTICAL TECHNOLOGY

Editor: Professor M. H. RUBINSTEIN, School of Health Sciences, Liverpool Polytechnic

UNDERSTANDING EXPERIMENTAL DESIGN AND INTERPRETATION IN PHARMACEUTICS
N. A. Armstrong & K. C. James
MICROBIAL QUALITY ASSURANCE IN PHARMACEUTICALS, COSMETICS AND TOILETRIES
Edited by S. Bloomfield *et al.*
DRUG DISCOVERY TECHNOLOGIES
C. Clark & W. H. Moos
PHARMACEUTICAL PRODUCTION FACILITIES: Design and Applications
G. Cole
PHARMACEUTICAL TABLET AND PELLET COATING
G. Cole
THE PHARMACY AND PHARMACOTHERAPY OF ASTHMA
Edited by P. F. D'Arcy & J. C. McElnay
GUIDE TO MICROBIOLOGICAL CONTROL IN PHARMACEUTICALS
Edited by S. P. Denyer & R. M. Baird
PHARMACEUTICAL THERMAL ANALYSIS: Techniques and Applications
J. L. Ford and P. Timmins
PHYSICO-CHEMICAL PROPERTIES OF DRUGS: A Handbook for Pharmaceutical Scientists
P. Gould
DRUG DELIVERY TO THE GASTROINTESTINAL TRACT
Edited by J. G. Hardy, S. S. Davis and C. G. Wilson
HANDBOOK OF PHARMACOKINETICS: Toxicity Assessment of Chemicals
J. P. Labaune
TABLET MACHINE INSTRUMENTATION IN PHARMACEUTICS: Principles and Practice
P. Ridgway Watt
PHARMACEUTICAL CHEMISTRY, Volume 1 Drug Synthesis
H. J. Roth *et al.*
PHARMACEUTICAL CHEMISTRY, Volume 2 Drug Analysis
H. J. Roth *et al.*
PHARMACEUTICAL TECHNOLOGY: Controlled Drug Release, Volume 1
Edited by M. H. Rubinstein
PHARMACEUTICAL TECHNOLOGY: Controlled Drug Release, Volume 2*
Edited by M. H. Rubinstein
PHARMACEUTICAL TECHNOLOGY: Tableting Technology, Volume 1
Edited by M. H. Rubinstein
PHARMACEUTICAL TECHNOLOGY: Tableting Technology, Volume 2*
Edited by M. H. Rubinstein
PHARMACEUTICAL TECHNOLOGY: Drug Stability
Edited by M. H. Rubinstein
PHARMACEUTICAL TECHNOLOGY: Drug Targeting*
Edited by M. H. Rubinstein
UNDERSTANDING ANTIBACTERIAL ACTION AND RESISTANCE
A. D. Russell and I. Chopra
RADIOPHARMACEUTICALS USING RADIOACTIVE COMPOUNDS IN PHARMACEUTICS AND MEDICINE
Edited by A. Theobald
PHARMACEUTICAL PREFORMULATION: The Physicochemical Properties of Drug Substances
J. I. Wells
PHYSIOLOGICAL PHARMACEUTICS: Biological Barriers to Drug Absorption
C. G. Wilson & N. Washington
PHARMACOKINETIC MODELLING USING STELLA ON THE APPLE™ MACINTOSH™
C. Washington, N. Washington & C. Wilson
* *In preparation*

UNDERSTANDING EXPERIMENTAL DESIGN AND INTERPRETATION IN PHARMACEUTICS

N. A. ARMSTRONG B.Pharm., Ph.D., F.R.Pharm.S., MCPP
K. C. JAMES M.Pharm., Ph.D., D.Sc., FRSC, F.R.Pharm.S., C.Chem.
Welsh School of Pharmacy
Division of Pharmaceutics, University of Wales, College of Cardiff

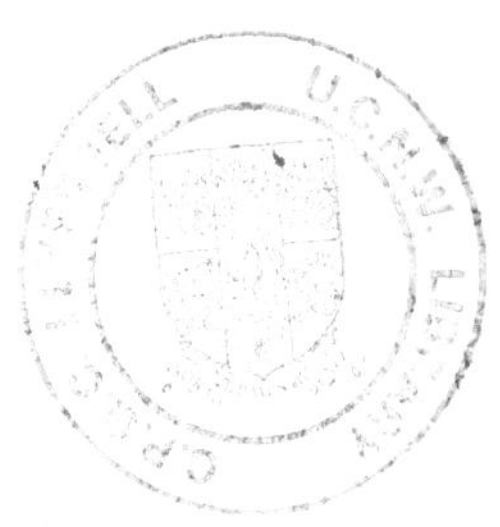

ELLIS HORWOOD
NEW YORK LONDON TORONTO SYDNEY TOKYO SINGAPORE

First published in 1990 by
ELLIS HORWOOD LIMITED
Market Cross House, Cooper Street,
Chichester, West Sussex, PO19 1EB, England

A division of
Simon & Schuster International Group
A Paramount Communications Company

Typeset in Times by Ellis Horwood Limited
Printed and bound in Great Britain
by Hartnolls, Bodmin, Cornwall

British Library Cataloguing in Publication Data

Armstrong, N. A.
Experimental design and interpretation in pharmaceutics.
1. Pharmaceutics
I. Title II. James, K. C.
615.19

ISBN 0–13–293465–5 (Library Edn.)
ISBN 0–13–931817–8 (Student Edn.)

Library of Congress Cataloging-in-Publication Data

Armstrong, N. A. (N. Anthony)
Understanding experimental design and interpretation in pharmaceutics / N. A. Armstrong, K. C. James.
p. cm. — (Ellis Horwood series in pharmaceutical technology)
Includes bibliographical references and index
ISBN 0–13–293465–5 (Library Edn.)
ISBN 0–13–931817–8 (Student Edn.)
1. Drugs — Design. 2. Experimental design. I. James, Kenneth C., 1924– . II. Title.
III. Series.
[DNLM: 1. Drug Design. 2. Technology, Pharmaceutical. QV 744 A737u]
RS420.A76 1990
615′.19–dc20 90–5118
CIP

Table of contents

1

Introduction to experimental design

Experimentation is expensive in terms of time, manpower and resources. It is therefore reasonable to ask if experimentation can be made more efficient, thereby reducing the waste of time and money.

Scientific principles of experimental design have been available for a considerable time. Much of the work originated with Sir Ronald Fisher and Professor Frank Yates. They worked together at the Rothamsted Agricultural Research Station, and there is an undeniably agricultural 'feel' to some of their terminology. The principles they and others devised have found application in a variety of fields, but it is surprising how little they have been used in pharmaceutical systems. The reasons for this neglect are a matter of speculation, but there is no doubt that the principles of experimental design do have a pharmaceutical applicability.

Attention is drawn to the title 'Experimental Design'. Experimentation can be defined as the investigation of a defined area with a firm objective, using appropriate tools and drawing conclusions which are justified by the experimental data so obtained.

Most experiments consist essentially of measuring the effect that one or more factors have on the outcome of the experiment. The factors are the independent variables and the outcome is the dependent variable.

The overall experimental process can be divided into a number of stages.

(1) Statement of the problem. What is the experiment supposed to achieve; what is its objective?
(2) The choice of factors to be investigated, and the levels of those factors which are to be used.
(3) The selection of a suitable response. This may be defined in the statement of the problem, stage (1), but it may not. If the latter, then we must be sure that the measurement of the chosen response will really contribute to achievement of the objective. The accuracy of the proposed methods of measuring the response must also be considered.
(4) The choice of the experimental design. This is often a balance between cost and

statistical validity. The more an experiment is replicated, the greater the reliability of the results. However, replication increases cost and the experimenter must therefore consider the acceptable degree of uncertainty. This in turn is governed by the number of replicates which can be afforded. Inextricably linked with this stage is selection of the method to be used to analyse the data.

(5) Performance of the experiment: the data collection process. This will follow the experimental design laid down earlier.
(6) Data analysis using methods defined earlier.
(7) Conclusions.

All too often, the objective of the experiment is imperfectly defined, the experiment is then carried out and only at that point are methods of data analysis considered. It is then discovered that the experimental design is deficient and has provided insufficient and/or inappropriate data for the most effective form of analysis to be carried out.

Thus the term 'experimental design' must include not only the proposed experimental methodology but also the methods whereby the data from the experiments are to be analysed. The importance of considering both parts of this definition together cannot be overemphasized.

A further point which must be considered at this stage is the availability of computing facilities, whether mainframe or PC. The advantages of the computer are obvious. The chore of often repetitive calculation has been removed, and so an undeniable disincentive to use statistical methods has been removed at the same time. However, using a computer can give rise to two related problems. The first is to make absolute reliance on the computer — if the computer says so, it must be so. The second is the assumption that the computer can take unreliable data or data from a badly designed experiment and somehow transform them into a result which can be relied upon. The computer jargon GIGO — garbage in, garbage out — is just as appropriate to problems of experimental design as to other areas in which computers are used.

USEFUL GENERAL REFERENCES

Bolton, S. (1984) *Pharmaceutical Statistics*, Marcel Dekker.

Burley, D. M. (1974) *Studies in Optimisation*, Intertext.

Finney, D. J. *An introduction to the theory of experimental design*, University of Chicago Press.

Fisher, R. A. (1937) *The Design of Experiments*, Oliver and Hall.

Hicks, C. R., (1982) *Fundamental Concepts in the Design of Experiments*, Holt-Saunders.

Montgomery, D. C. (1976) *Design and Analysis of Experiments*, Wiley.

2

Comparison of mean values

A common feature of many experimental programmes is to obtain groups of data under two or more sets of experimental conditions. The question then arises 'Has the change in experimental conditions affected the data?' The question may be rephrased to a more precise form: 'Do the means of each group differ significantly, or are all groups really taken from the same population, the change in experimental conditions having had no significant effect?'

A variety of experimental techniques exists to answer this question. Hence it is all too easy to select an inappropriate technique with misleading results.

For selection of the correct procedure, a number of further questions must be asked.

(1) Are there more than two sets of data? If so, analysis of variance should be used.
(2) If there are only two sets of data, do these sets represent the total population or do they represent samples drawn from a larger population? In other words, do we know the variance of the whole population? Examples of the former could be sets of examination results, when the performance of every candidate is known. Also in long-running industrial process, where many batches have been made under identical conditions, the pooled variance of all the batches will be very close to the variance of the total population or universe. In such cases, a statistical test using the normal standard deviate is used.

If the variance of the whole population is not known or cannot be inferred, then the Student's *t* test is usually employed.

Thus the position can be summarized by the following scheme.

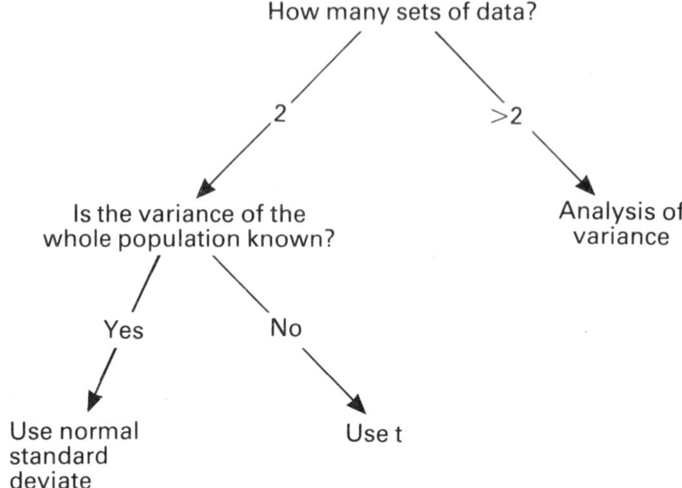

COMPARISON OF TWO MEANS WHEN THE VARIANCE OF THE WHOLE POPULATION IS KNOWN

This is best illustrated by a simple example.

Undergraduate students are taught a particular class in groups of 10. There are 2 such groups, designated A and B. At the end of the course, both groups are given the same test. Based on their performance in the test, the students in group B assert that their teacher is incompetent. The examination marks are given in Table 2.1.

Table 2.1 — Marks gained by two groups of students (%)

	Group A	Group B
	52	55
	59	45
	60	66
	44	42
	70	56
	54	45
	44	44
	56	48
	56	53
	51	38
Mean	54.6	49.2
Standard deviation	7.7	8.3

The means show a difference of over 5 on marks around 50, so this difference seems quite large. Nevertheless, there is considerable scatter within each group.

The procedure here is to construct confidence intervals for the means.

The confidence interval for the mean of group A is given by equation (2.1)

$$\text{Confidence interval} = \bar{x}_A + \frac{Z_P\sigma}{\sqrt{N_A}} \tag{2.1}$$

where $\bar{x}_A$ is the mean of group A, σ is the standard deviation of group A, N_A is the number of observations in group A, P is the required level of probability, Z_P is the normal deviate.

A key point to grasp here is that the required level of probability must be selected before the calculation can be made. Since the accusation is fundamentally one of incompetence, with potentially severe consequences for the person involved, a high level of probability is appropriate. For the sake of argument, a value of P of 0.99 is selected.

The value of Z to choose from Appendix 2.1 is obtained from the column headed by a value equal to $(P+1)/2$, in this case 0.995. Thus $Z = 2.576$.

Hence the confidence limits for the mean of group A is

$$54.6 \pm \frac{2.576 \times 7.7}{\sqrt{10}}$$
$$= 54.6 \pm 6.3$$
$$= 48.3 \text{ to } 60.9$$

The mean of group B comes within this range. Thus it can be claimed that there is no significant difference between the means at a probability level of 0.99. If a probability level of 0.95 had been selected, then there would have been a significant difference between the means and an explanation would be sought. Even so, the incompetence of the tutor is not the only possible explanation.

This example also shows the influence of the number of observations in the group. Thus if group A had contained only five students, but with the same mean and standard deviation, then the confidence limits would be considerably widened at 45.7 to 63.5.

COMPARISON OF TWO MEANS WHEN THE VARIANCE OF THE WHOLE POPULATION IS NOT KNOWN

This situation arises when the experimental data are obtained from samples taken from a much larger population. Thus unless every member of the population were to be tested, then the variance must be estimated from data obtained from samples.

The test statistic t is calculated from equation (2.2)

$$t = \frac{\bar{x}_A - \bar{x}_B}{\sqrt{\dfrac{s_A}{n_A} + \dfrac{s_B}{n_B}}} \tag{2.2}$$

where $\bar{x}_A$ and $\bar{x}_B$, s_A and s_B, and n_A and n_B are the mean, standard deviation and number of data points in groups A and B respectively. As an example, consider the following situation.

Batches of capsules are prepared by two processes, A and B. Samples of 6 capsules from each batch are subjected to the disintegration test of the BP. Does a significant difference exist between the means of the two batches? The data are given in Table. 2.2.

Table 2.2 — Disintegration time of six capsules (minutes) taken from batches A and B

	Batch A	Batch B
	11.1	6.2
	10.3	7.3
	13.0	8.2
	14.3	8.3
	11.2	7.5
	14.7	6.5
Mean	12.43	7.33
Variance	3.35	0.74
SD	1.83	0.86
n	6	6

From equation (2.2), $t = 6.17$. Reference to tabulated values of t show that for there to be a significant difference with 10 degrees of freedom at $P = 0.01$, t should not be less than 3.17. Hence the difference between the means is significant at that level.

The formula used to calculate t in equation (2.2) is that used when the variances of the two populations differ considerably. In this case, the ratio between the variances is $3.35/0.74 = 4.52$.

If the variances are closer than this (a ratio of less than 3 is a good rule of thumb), then equation (2.3) may be used instead.

$$t = \frac{\bar{x}_A - \bar{x}_B}{s\sqrt{\frac{1}{n_A} + \frac{1}{n_B}}} \tag{2.3}$$

Use of equation (2.2) gives a more conservative estimate of significance than equation (2.3), even when both samples have similar variances. Hence it is probably prudent to use equation (2.2).

COMPARISON OF MEANS AMONG MORE THAN TWO GROUPS OF DATA

The examples discussed so far involve the comparison of the means of only two groups of data. However, it may be that there are more than two groups. Consider the following example.

Tablets are made using three different formulations, A, B and C. A sample of ten tablets is selected from each batch and the crushing strength of each tablet measured. The data are given in Table 2.3. Do the mean tablets strengths differ significantly?

Table 2.3 — The crushing strengths of tablets (kg) from batches A, B and C

	Batch A	Batch B	Batch C
	5.2	5.5	3.8
	5.9	4.5	4.8
	6.0	6.6	5.1
	4.4	4.2	4.2
	7.0	5.6	3.3
	5.4	4.5	3.5
	4.4	4.4	4.0
	5.6	4.8	1.7
	5.6	5.3	5.9
	5.1	3.8	4.8
Total	54.6	49.2	41.1
Mean	5.46	4.92	4.11
Grand total	144.9		
Grand mean	4.83		

A possible way forward would be to carry out multiple t-tests, i.e. compare batch A with batch B, batch B with batch C and batch C with batch A.

The results of this are:

Batches A and B, $t = 1.51$
Batches B and C, $t = 1.79$
Batchcs A and C, $t = 3.06$

Thus Batch C is significantly lower than Batch A at a probability level of 95%.

There is a serious flaw in this approach. A probability level of 95% means that in 95% of cases the statement associated with that level will be correct. In other words, in 5% of cases it will be wrong. Now three probability statements have been made, and if there is a 5% chance of each being wrong, it follows that there is a 15% chance of one of the three being wrong. Furthermore there is no way of knowing which result is incorrect. Thus, as the number of groups of data increases, there is a rapidly diminishing chance of a correct overall assessment being made using the t-test.

The proper way to proceed in these circumstances is to use analysis of variance.

ANALYSIS OF VARIANCE

Analysis of variance (ANOVA) is a most powerful statistical tool. It permits the comparison of the means of several populations. It assumes that a random sample has been taken from each population, that each population has a normal distribution and that all the populations have the same variance (in practice, the last two requirements are not essential if sample sizes are approximately equal). The question that analysis of variance seeks to answer is, 'Do all the populations have the same mean?'

Obviously, within each group of data there will be scatter, and there will also be scatter between groups. The variation within a group is unexplained variation, arising from random differences between the subjects and sources of variation which are either unknown or are being ignored. The problem is to answer the question 'Is the between-group variation significantly greater than the within-group variation?'

The ANOVA procedure is best approached as a series of numbered steps, using as an example the data given in Table 2.3.

(1) Calculate the total and the mean of every column.

(2) Calculate the grand total and the grand mean. (The results of these first two steps already appear in Table 2.3.)

(3) Calculate the (grand total)^2/(number of observations):

$$= (144.9)^2/30 = 699.87$$

This term is used several times in this calculation. It is often called the 'correction term' and denoted by the letter C.

(4) Calculate the sum of (every result)^2

$$= 5.2^2 + 5.9^2 + \ldots + 4.8^2$$
$$= 732.71$$

(5) Subtract C from the result of step 4

$$= 732.71 - 699.87$$
$$= 32.84$$

This gives the value of the term $(\Sigma x^2 - (\Sigma x)^2/N)$, and is known as the total sum of squares.

(6) Calculate the sums of squares between means.

$$= \frac{54.6^2}{10} + \frac{49.2^2}{10} + \frac{41.1^2}{10} - C$$
$$= (298.12 + 240.06 + 168.92) - 699.87$$
$$= 9.23.$$

(7) Calculate the difference between the total sum of squares and the sum of squares between means.

$$= 32.84 - 9.23$$
$$= 23.61$$

This is known as the residual sum of squares.

(8) At this stage it is useful to draw up an analysis of variance table (Table 2.4).

The degrees of freedom for the whole experiment are $(3 \times 10) - 1 = 29$. There are three groups of tablets and hence three means. There are hence

Table 2.4 — ANOVA table of data from Table 2.3

Source of error	Sum of squares	Degrees of freedom	Mean square	F
Total	32.84	29	—	—
Between means	9.23	2	—	—
Within each group	23.61	27	—	—

(3 – 1) = 2 degrees of freedom here. Thus the residual sum of squares has (29 – 2) = 27 degrees of freedom.

(9) The mean squares are obtained by dividing the sum of squares by the relevant number of degrees of freedom. The two means squares are thus 4.62 and 0.87. These are added to Table 2.4.

(10) The F ratio (named after Fisher) is the ratio between the mean squares. This equals 5.31 and is added to Table 2.4.

(11) The ANOVA table is now complete (Table 2.5).

Table 2.5 — Complete ANOVA table of tablet strength data from Table 2.3

Source of error	Sum of squares	Degrees of freedom	Mean square	F
Total	32.84	29	—	—
Between means	9.23	2	4.62	5.31
Within each group	23.61	27	0.87	

(12) The ratio is compared with appropriate tabulated values of F. Separate F tables are given for each probability level. Use of the table requires two values for degrees of freedom. That for the 'mean square between means' forms the top row of the table, and the mean square of the residuals the left-hand column of the table.

For the data under consideration, and using a significance level of 0.05, then the tabulated value of F is 3.35. (Remember there are 2 degrees of freedom between treatments and 27 degrees of freedom within treatments. The first of these is on the first row of the table and the second is on the left-hand column of the table.)

Thus there is a significant difference between the means at a probability level of 95%. The corresponding value for F at a probability level of 99% is 5.49. This is greater than the calculated value and so the difference is not significant at a probability of 99%.

The value of analysis of variance as a tool should now be apparent. There is no limit to the number of groups of data, and all groups need not necessarily be the same size.

Analysis of variance shows that a significant difference occurs between the means of a number of groups of data. It gives no information as to which group is significantly different from the others. Therefore, having established that there are differences, it is necessary to establish which groups differ. Do they all differ from each other or are there some which are effectively the same?

There are a number of tests available which help to establish this point. The simplest of these is to calculate the Least Significant Difference.

LEAST SIGNIFICANT DIFFERENCE

This test uses Student's t. It will be recalled that it was shown earlier that this was an inappropriate test to use when there were more than two groups of data to **establish** whether a significant difference exists. However, it will now be used **after** a significant difference has been shown to exist by analysis of variance.

$$\text{Since } t = \frac{\bar{x}_A - \bar{x}_B}{\sqrt{s^2\left(\frac{1}{n_A} + \frac{1}{n_B}\right)}}$$

then the least significant difference is

$$\bar{x}_A - \bar{x}_B = t\sqrt{S^2(1/n_A + 1/n_B)}$$

where t is the tabulated value of t with the appropriate number of degrees of freedom and required significance level (in this case 2.06).

S^2, the variance, is equal to the mean square within each group (in this case 0.87). Therefore the least significant difference

$$= 2.06 \times \sqrt{0.87 \times 2/10}$$
$$= 0.85.$$

The differences between the means are;

A − B = 0.54
B − C = 0.81
A − C = 1.35

Thus any difference above 0.85 is significant, and in this case, the difference between A and C proves significant. Also though not significant, the difference between B and C approaches 0.85. Hence this is a reasonable indication that, of the three treatments, group C is the one which is most likely to be different.

There are several other methods of determining which, if any, treatment gives significantly different results after analysis of variance. These include the Duncan multiple range test, the Dunnett test, the Tukey multiple range test and the Scheffe test. All give a parameter equivalent to the least significant difference and each has its own claimed advantages. Interested readers should refer to the associated bibliography.

TWO-WAY ANALYSIS OF VARIANCE

The analysis of variance test described earlier is more properly called 'one-way analysis of variance'. One factor is deliberately changed (e.g. Formulation A, B or C). However, a situation may arise when two factors are changed — for example, results may be obtained on different equipment or in different geographical areas. The aim is therefore to determine whether the treatments have a significantly different effect while taking the known underlying variation into account. Two-way analysis of variance is employed in this case.

The situation is best illustrated by a worked example.

A multinational pharmaceutical company produces tablets containing a certain drug in three different countries. Each country uses its own formulation for the tablets. It is decided to produce the tablets using the same formulation in all three countries. *In vitro* dissolution data appear to indicate differences among the three formulations, but the differences might be due to the fact that the formulations are produced at different sites.

Let the formulations be designated A, B and C and the three sites of manufacture I, II and III. The obvious way forward is to produce batches at all three sites using all three formulations. Three batches of each formulation would be obtained and an analysis of variance would show whether significant differences between the batches were present. However, there might be geographical factors which affect the results such as equipment, personnel or the familiarity a particular site will have with the production of its local formulation. In fact apparent differences between formulations might be almost entirely due to such factors.

The following experiments are therefore carried out. Each formulation is prepared at each site and the dissolution data obtained. The data are given in Table 2.6.

Table 2.6 — Time for 50% of drug to be dissolved (min.)

	Formulation			Site
	A	B	C	total
Site				
I	35	26	23	84
II	41	29	26	96
III	42	29	19	90
Formulation total	118	84	68	270

The total variance is made up of three components, namely the variance among formulations, the variance among sites of manufacture and the residual variance.

The stages in the calculation of two-way analysis of variance are very similar to those in a one-way analysis.

Thus

(1) Calculate the grand total, i.e. the sum of all the data $=35+26+23 \ldots +19=270$.
(2) Calculate the totals for each site and for each formulation, e.g. Site I $=35+26+23=84$, etc.
(3) Calculate the correction term

$$= \frac{\text{grand total}^2}{\text{number of observations}}$$
$$= 270^2/9 = 8100$$

(4) Calculate the total sum of squares

$$= (35^2 + 26^2 + \ \ldots \ + 19^2) - 8100$$
$$= 494.$$

(5) Calculate the 'between formulations' sum of squares

$$= 1/3(118^2 + 84^2 + 68^2) - 8100$$
$$= 434.7.$$

(6) Calculate the 'between sites' sum of squares

$$= 1/3\ (84^2 + 96^2 + 90^2) - 8100$$
$$= 24$$

(7) Construct an analysis of variance table (Table 2.7).

Table 2.7 — Analysis of variance table of data from Table 2.6

Source	Sum of squares	Degrees of freedom	Mean square	F
Total	494	8	—	—
Between formulations	434.7	2	217.5	24.7
Between sites	24.0	2	12.0	1.4
Residual	35.3	4	8.8	—

The degrees of freedom are calculated as follows. The total number of degrees of freedom for N observations are $(N-1)$, in this case 8. If there are R rows in the table and C columns, then the numbers of degrees of freedom associated with rows and columns are $(R-1)$ and $(C-1)$ respectively. In this case, there are 2 degrees of freedom associated with both. The degrees of freedom associated with the residuals are $(R-1)\times(C-1)$, in this case 4.

The tabulated value of F at a 95% level of significance is 6.94. Thus there appears to be a significant difference among the formulations, but not among the sites.

The above example is perhaps unrealistic in that only single results are reported from each site. Though these could be means of perhaps six individual tests, it is likely that the whole experiment would be repeated on each site.

Assume that duplication has taken place, and the individual data points are as shown in Table 2.8.

Table 2.8 — Time for 50% of drug to be dissolved (min). Duplicated data

	Formulation			Site total
	A	B	C	
Site				
I	33, 37	22, 30	23, 23	168
II	40, 42	24, 34	24, 28	192
III	40, 44	26, 32	19, 19	180
Formulation total	236	168	136	540

(1) Calculate the correction term

$$= 540^2/18$$
$$= 16200$$

(2) Calculate the total sum of squares

$$= (33^2 + 37^2 + \ \dots\ 19^2) - 16200$$
$$= 1114.$$

(3) Calculate the 'formulation' sum of squares

$$= 1/6(236^2 + 168^2 + 136^2) - 16200$$
$$= 869.$$

(4) Calculate the 'site' sum of squares

$$= 1/6(168^2 + 192^2 + 180^2) - 16200$$
$$= 48.$$

(5) An additional term in this calculation is the 'within cell' sum of squares

$$= 1/2[(33-37)^2 + (22-30)^2 + \dots + (19-19)^2] - 16200$$
$$= 126$$

(6) Calculate the residual sum of squares

$$= [1114 - (869 + 48 + 126)]$$
$$= 71.$$

The total number of degrees of freedom are 17. As before, there are two degrees of freedom associated with formulation and sites, and four associated with residuals. Degrees of freedom associated with the error within cells therefore number 9.

Table 2.9 — Analysis of variance table of data from Table 2.8

Source	Sum of squares	Degrees of freedom	Mean square	*F*
Total	1114	17	—	—
Between formulations	869	2	434.5	55.0
Between sites	48	2	24.0	3.0
Residual	126	4	31.5	4.0
Within cells	71	9	7.9	—

The ANOVA table can be constructed (Table 2.9) and, in this case, *F* is calculated by dividing the mean squares for formulation, site and residuals by the mean square for 'within cells'.

3

Non-parametric treatments

Many of the tests so far employed, such as *t* and ANOVA, depend on the assumption that the populations involved are normally distributed. In many cases this cannot be known with certainty, though it can often be assumed. It should also be borne in mind that the distribution of a sample mean approaches that of a normal distribution as the sample size is increased. However, increase in the sample size may not be practicable.

A further consideration is that the data to be manipulated by parameteric methods must have numerical values. Ordinal data based on rank order, e.g. social class, severity of reaction, are not amenable to parametric treatment.

There is a series of non-parametric tests available which are designed to handle such information. These have the distinction of making no prior assumptions of the underlying distribution and parameters of the population.

As in parametric tests of comparison, the distinction must be made as to whether the two samples come from independent populations or whether the variates are paired in some way, perhaps by each subject acting as its own control. This obviously depends on the design of the experiment. Hence here is another example of the design of the experiment and the method of evaluating the results being inextricably linked.

Some of the tests which can be used are:

(1) Sign test for paired data.
(2) Wilcoxon signed rank test for paired data (the Mann–Whitney U-test is very similar).
(3) The Wilcoxon two-sample test for unpaired data.

NON-PARAMETRIC TESTS FOR PAIRED DATA

The sign test

This is used to test the significance of the difference between the means of two sets of data in a paired experiment. Each subject thus acts as its own control. Only the sign of the differences between each pair of data points is used, and, because of its simplicity, this test may be used for a rapid examination of data before a more sensitive test is applied.

Consider the following example. The dissolution rate of tablets is measured on a well-defined piece of apparatus. It is suggested that certain modifications will improve the apparatus. A number of different tablet formulations are tested on both types of apparatus, giving the results in Table 3.1.

Table 3.1 — Percentage of active ingredient dissolved in 30 minutes (%)

Tablet formulation	Apparatus I	Apparatus II	Difference (II–I)
A	83	88	+5
B	59	66	+7
C	78	83	+5
D	79	79	0
E	88	92	+4
F	82	90	+8
G	90	92	+2
H	81	83	+2
I	87	77	−10
J	65	68	+3
K	68	72	+4
L	83	89	+6

Tabulation of the differences between Apparatus I and Apparatus II shows ten positive signs, one negative sign and, in one case, both pieces of apparatus give the same result. If the two pieces of apparatus were truly equivalent, then the probability of a '+' or a '−' for any given formulation would be 0.5. When the number of observations is small, the probabilities of various experimental outcomes can be calculated from the binomial distribution. The number of positive or negative signs needed for significance for the sign test is given in Table 3.2. For larger samples, equation (3.1) is used.

$$Z = \frac{|\text{number of pluses} - \text{number of minuses}| - 0.5}{\sqrt{(\text{number of pluses} + \text{number of minuses})}} \tag{3.1}$$

If Z is greater than 1.96, there is a significant difference at the 5% level of probability, and if greater than 2.60, a significant difference at the 1% level.

For the purposes of these calculations, results which are tied are ignored. From Table 3.2, it is seen that for 11 pairs of observations, there should be at least 10 with the same sign for a significant difference at the 5% level, and so the two pieces of dissolution apparatus appear to differ significantly at this level. For a significant

Table 3.2 — Number of positive or negative signs needed for significance for the sign test

Sample size	Number of positive or negative signs	
	5%	1%
6	6	—
7	7	—
8	8	8
9	8	9
10	9	10
11	10	11
12	10	11
13	11	12
14	12	13
15	12	13
16	13	14
17	13	15
18	14	15
19	15	16
20	15	17

difference at the 1% level, all 11 pairs should have the same sign, which does not apply in the present case.

A further important point arises from Table 3.2. For a 5% level of significance, the smallest sample size which can be expected to yield a significant result is 6, and for a 1% level, the corresponding sample size is 8. It follows therefore that if this test is to be used, planning of the experiment must take these required sample sizes into account.

THE WILCOXON SIGNED RANK TEST

This is a more sensitive non-parametric test. In this the magnitude of the difference between the paired variates as well as its sign is taken into account.

Using the same data as given in Table 3.1, the differences are ranked in order of increasing magnitude disregarding the sign. Identical values are given a mean rank and ties are discounted (Table 3.3a).

The results are then rearranged taking into account the signs and their magnitude. Ranks with negative signs and ranks with positive signs are summed separately (Table 3.3b).

Table 3.4 gives the values of the smaller of the two rank sums at a 5% significance level for a range of sample sizes. The smaller rank sum must be equal to or less than the number given in the table.

As before, significance is established at the 5% level but not at the 1% level.

Table 3.3a — Assigned ranks with and without signs for data from Table 3.1

Formation	Assigned rank	Rank with sign
G	1.5	+1.5
H	1.5	+1.5
J	3	+3
E	4.5	+4.5
K	4.5	+4.5
A	6.5	+6.5
C	6.5	+6.5
L	8	−8
B	9	+9
F	10	+10
I	11	+11

Table 3.3b — Ranks with positive and negative signs derived from Table 3.3a

Ranks with positive signs	Ranks with negative signs
+1.5	−8
+1.5	—
+3	—
+4.5	—
+4.5	—
+6.5	—
+6.5	—
+9	—
+10	—
+11	—
sum +61	−8

Table 3.4 — Values giving significance for the Wilcoxon signed rank test

Sample size	5% significance	1% significance
6	0	—
7	2	—
8	3	0
9	5	1
10	8	3
11	10	5
12	13	7
13	17	10
14	21	13
15	25	16
16	30	19
17	35	23
18	40	28
19	46	32
20	52	37

NON-PARAMETRIC TESTS FOR UNPAIRED DATA

The Wilcoxon two-sample test

This test deals with two groups of data which have been obtained independently, i.e. an item in one group does not act as the control for an item in the second group. The data need not be normally distributed, and the groups need not even be of the same size. As before, the test is best illustrated by an example.

Armstrong *et al.* (1988) described a method for measuring the release of drugs from oily bases. The drug was released into an aqueous medium, which was then analysed. The method was used to compare drug release from bases of differing composition, and some of the data obtained are given in Table 3.5.

The task is to answer the question 'Does a change of base have a significant effect on drug release?'

Cursory examination of the data indicates that Base B gives a slower release in that most of the values for Base B are less than those of Base A. It would seem that a *t*-test may be appropriate but there is little evidence that the data are normally distributed.

The Wilcoxon two-sample test is carried out by arranging the data in ascending order of magnitude (Table 3.6).

Table 3.5 — Drug release from two topical bases after 120 minutes (data are mg% in the aqueous phase)

	Base A	Base B
	0.782	0.742
	0.790	0.779
	0.798	0.748
	0.772	0.764
	0.790	0.757
Mean	0.786	0.758

Table 3.6 — Drug release from two topical bases after 120 minutes (data are ranked in increasing order of magnitude)

Base A	Rank	Base B	Rank
0.782	7	0.742	1
0.790	8.5	0.779	6
0.798	10	0.748	2
0.772	5	0.764	4
0.790	8.5	0.757	3

The sum of the ranks of data from Base B is

$$1+2+3+4+6=16$$

Similarly, for Base A, the sum is

$$5+7+8.5+8.5+10=39$$

Adding 16 to 39 gives 55, which is the sum of the integers 1 to 10. This has no bearing on the outcome of the experiment but it serves as a useful check that the ordering has been carried out correctly.

Note that in Group A there are two identical results (0.790). If these were slightly different, they would be ranked 8 and 9 in the ascending order. These positions are therefore averaged (8.5) and this rank given to each. The total remains the same.

If there were no difference between drug release for the two bases, then the totals for each group would be about the same. The difference (16 to 39) looks large, and is an indication of a difference in drug release, but nevertheless could have occurred by chance.

The next step is to determine how many of all the possible arrangements of the numbers 1 to 10 will give a total of 16 of less.

These are two, namely

$$1+2+3+4+5=15$$

$$1+2+3+4+6=16$$

The number of possible combinations of ten objects taking five at a time is given by the formula $10!/[5!-(10-5)!]=252$.

Thus the probability of obtaining a sum of ranks less than or equal to 16 is given by

$$P=\frac{2}{252}=0.00794$$

P, as calculated above, is for a one-tail test, i.e. it is the probability that release from Base B is less than that from Base A. If a significant difference between Base A and Base B is to be established, then P is doubled and becomes 0.0159. From the above, it follows that a significant difference occurs between the two bases at a probability level of 1.59%.

It is of interest to apply the t-test to these data, assuming for the moment that both populations follow a normal distribution. t is found to be 3.63, which is equivalent to a probability of 1.66% for a two-tail test, close to that calculated by the Wilcoxon test.

REFERENCE

Armstrong, N. A., Griffiths, H-A. and James, K. C. (1988) An *in-vitro* model to simulate drug release from oily media, *Int. J. Pharmaceut.* **41** 115–119.

4

Factorial design of experiments

A classical approach to experimentation is to investigate the effects of one experimental variable while keeping all others constant. This is a valid approach when the underlying laws relating cause and effect are known with some certainty. However, in many cases such knowledge is not available and it is not known which, out of the many variables which might affect the outcome of an experiments will prove the most important and hence justify more extensive study.

Furthermore it is possible that variables may interact with each other. Thus the magnitude of the effect caused by altering one factor will depend on the magnitude of one or more other factors. An experimental design which investigates one factor while keeping all others constant is unlikely to disclose the presence of such interactions.

Factorially designed experiments are intended to avoid such problems. It is a system of experimental design by which the factors involved in a reaction or a process can be identified and their relative importance assessed. It is thus a means of separating those factors which are important from those which are not.

TWO-LEVEL, TWO-FACTOR EXPERIMENTAL DESIGNS

The simplest factorial design is one in which two factors are studied at two levels, low and high.

Notation in factorially designed experiments

This can often be the source of confusion since a variety of conventions has been used. For example, the factor may be designated by an appropriate letter, e.g. T for temperature. The lower level is represented by a lower case t and the higher level by an upper case T. Alternatively, the two levels can be designated $T1$ and $T2$.

The most frequently encountered notation for two-level studies, and the one that will be adopted here, is to designate the factors by upper case letters, beginning with A. Any experiment in which factor A is at a high level is designated by the corresponding lower case letter, a. An experiment where factors A and B are at a higher level is thus designated *ab* and the experiment where all factors are at their lower level is denoted by (1). It is also useful to draw up a table of experiments in which low levels are shown by – and higher levels by +.

There is also a convention in the order in which the experiments are written down in the table, namely (1), *a*, *b*, *ab*. Thus the first row in the table denotes the experiment in which all factors are at their lower level, and the fourth is when factors *A* and *B* are at their higher levels. This convention, known as the standard order, is relatively unimportant with simple experiments, but its usefulness will become more apparent later. These points are illustrated in the following example.

Compound X is an ester. As such it would be expected to be hydrolysed when in aqueous solution. It is anticipated that the rate of hydrolysis would be influenced by temperature and that the reaction can be catalysed, perhaps by the presence of H^+ or OH^- ions.

A simple factorially designed experiment will help assess the relative importance of these two factors.

The procedure is as follows. Temperature is designated as factor *A* and two temperatures are selected. The second factor, *B* is the presence or absence of the catalyst. A table of experiments is then drawn up in the standard order (Table 4.1) so that four sets of experimental conditions are obtained.

Table 4.1 — A two-factor, two-level factorial design

Experiment	Factor A (temperature)	Factor B (catalyst)	Loss of X (%)
(1)	–	–	10
a	+	–	25
b	–	+	30
ab	+	+	45

Thus in experiment (1) the temperature is low and the catalyst is absent, whereas in experiment *ab* the solution contains the catalyst and is kept at an elevated temperature. The four experiments set up as described above are run for a specified time and the loss of drug X is measured, giving the results shown in the last column of Table 4.1.

The effect of the two factors can now be calculated. The effect of any given factor is the change in response produced by altering the level of that factor, averaged over the levels of all the other factors.

The combination of experimental variables and the results can be envisaged in the form of a square (Fig. 4.1). The effect of temperature is the mean of the results on the right-hand side of the square minus the mean of those on the left-hand side. Similarly, the effect of the catalyst is the average of all results on the top of the square minus that of those results on the bottom.

Thus the effect of factor *A* is

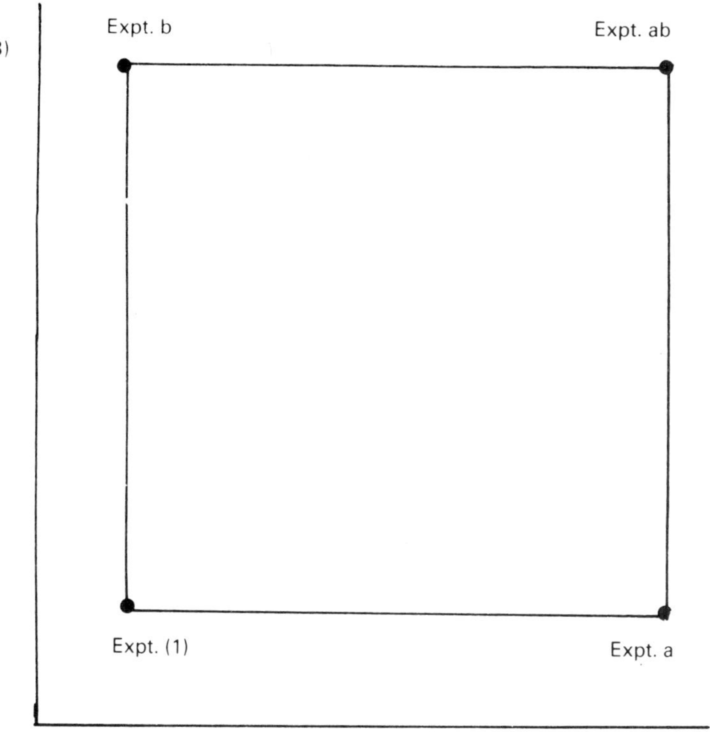

Fig. 4.1 — Representation of a two-factor, two-level experimental design.

$$\begin{aligned}&1/2\,[ab+a]-[b+(1)]\\&=1/2[ab+a-b-(1)]\\&=1/2[45+25-30-10]\\&=15.\end{aligned}$$

Similarly, the effect of factor B is

$$\begin{aligned}&1/2[ab+b]-[a+(1)]\\&=1/2[ab+b-a-(1)]\\&=1/2[45+30-25-10]\\&=20.\end{aligned}$$

Also the effect of factor A can be calculated by adding together all results in rows with a '+' in the factor A column, taking the mean and subtracting from it the mean of all those from rows with a '−' in that column. All these methods of calculation are essentially the same and give identical results.

Thus in this example both factors have an approximately equal effect and are therefore worthy of equal consideration.

However, consider an equally feasible alternative situation (Table 4.2).

By the same method of calculation, the effect of temperature is 10 and the effect of the catalyst is 60. In this case, the catalyst proves to be much the more important effect and attention should be concentrated on that.

The foregoing is a very straightforward example. However, the same principles can be used for much more complex systems.

FACTORIAL DESIGN WITH INTERACTION BETWEEN FACTORS

In the foregoing example an assumption has been made that the factors act independently to produce their effects. In many cases this will be so, but in others the level of one factor may govern the magnitude of the effect of another. This is termed factor interaction.

Interactions can often be detected graphically. In the data given in Table 4.1 an increase in temperature causes an additional loss of X of 15% (25%–10%), and the presence of a catalyst a loss of 20% (30%–10%). When both factors are at a high level, the total additonal loss is 35%, which is numerically equal to the total of the losses caused by the factors individually. Thus there is no interaction between the two factors. This situation is shown in Fig. 4.2. If a line is drawn joining the two results with factor B at a high level (experiments b and ab) and another line joining the two experiments in which factor B is at a low level (experiments (1) and a), then if no interaction occurs, two parallel lines will result.

A quantitative estimation of factor interaction is made as follows. A further column is added to Table 4.1. The signs in that column relating to the individual experiments are derived by the normal algebraic rules. Table 4.1 now becomes Table 4.3.

The magnitude of the interaction term is then calculated in the same way as that of the main factors, i.e. the mean of the results of all experiments with a + in the interaction column minus the mean of all those with a − in that column

$$\begin{aligned} &= 1/2[(1)+ab]-[a+b] \\ &= 1/2(10+45)-(25+30) \\ &= 0. \end{aligned}$$

If the combined effect of the two factors had been to produce a loss in X greater than that produced by the factors individually, then the interaction is said to be synergistic. An interaction which produced a decrease is antagonistic. In both cases, two parallel lines would not have been obtained in Fig. 4.2.

An example of how factors can interact is shown in the following example. A study is to be carried out to ascertain if aspirin tablets are stable in PVC controlled-dosage blister packs for use in nursing homes. As a control, identical tablets are packed in glass bottles closed by screw caps. The effects of temperature (designated factor A) and humidity (designated factor B) are to be studied, and a two-factor, two-level study is set up. Two temperatures (25°C and 45°C) and two relative humidities (50% and 80%) are selected. Aspirin stability is assessed by measuring the salicylic acid content of the tablets (in parts per million) after six months. The experimental conditions and analytical results are shown in Table 4.4.

Table 4.2 — A two-factor, two-level factorial design

Experiment	Factor A (temperature)	Factor B (catalyst)	Loss of X (%)
(1)	−	−	20
a	+	−	30
b	−	+	80
ab	+	+	90

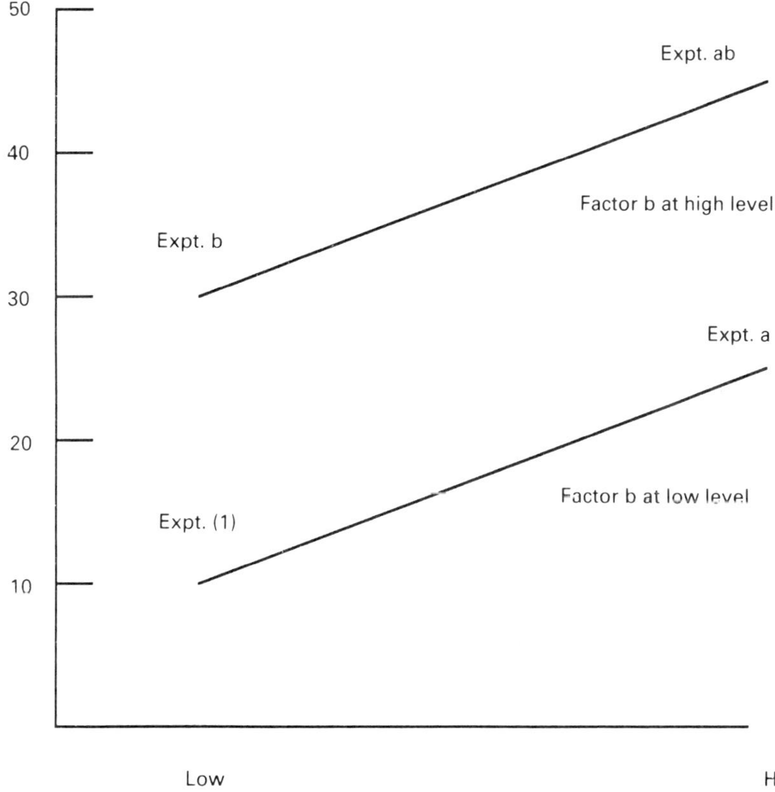

Fig. 4.2 — Two-factor, two-level design with no interaction.

Table 4.3 — Two-factor, two-level factorial design

Experiment	Factor A (temperature)	Factor B (catalyst)	Interaction of A and B	Loss of X (%)
(1)	−	−	+	10
a	+	−	−	25
b	−	+	−	30
ab	+	+	+	45

Table 4.4. — Factorially designed experiment to determine the stability of aspirin tablets

Experiment	Factor A (temperature)	Factor B (humidity)	Interaction of A and B	Salicylic acid content (ppm)	
				PVC	*Glass*
(1)	−	−	+	5.0	2.4
a	+	−	−	5.0	17.4
b	−	+	−	5.8	2.3
ab	+	+	+	32.1	20.0

A graphical treatment of the data is shown in Fig. 4.3, and there is clearly an interaction between the two factors when PVC is the packaging material. On the other hand, the two lines representing data from glass packaging are more or less parallel.

The magnitudes of the effects of temperature and humidity, and that of the interaction between them, are calculated as described above and are given in Table 4.5.

Thus for the PVC packs, the two factors and the interaction are all equally important, whereas for the glass pack the effect of humidity and the interaction are both negligible. Unlike the PVC pack, the glass pack is impermeable to water vapour. The interaction between the factors in the case of the PVC pack arises from the fact that the reaction involved, being hydrolytic, consumes water. When the external environment is of high humidity, water vapour can diffuse into the pack and 'feed' the reaction. With low external humidity, the concentration gradient of water is reversed, water vapour diffuses out of the pack and hence the reaction slows down. The tablets packed in glass are, as far as their environment is concerned, in conditions of constant humidity, and thus the reaction rate is governed solely by temperature.

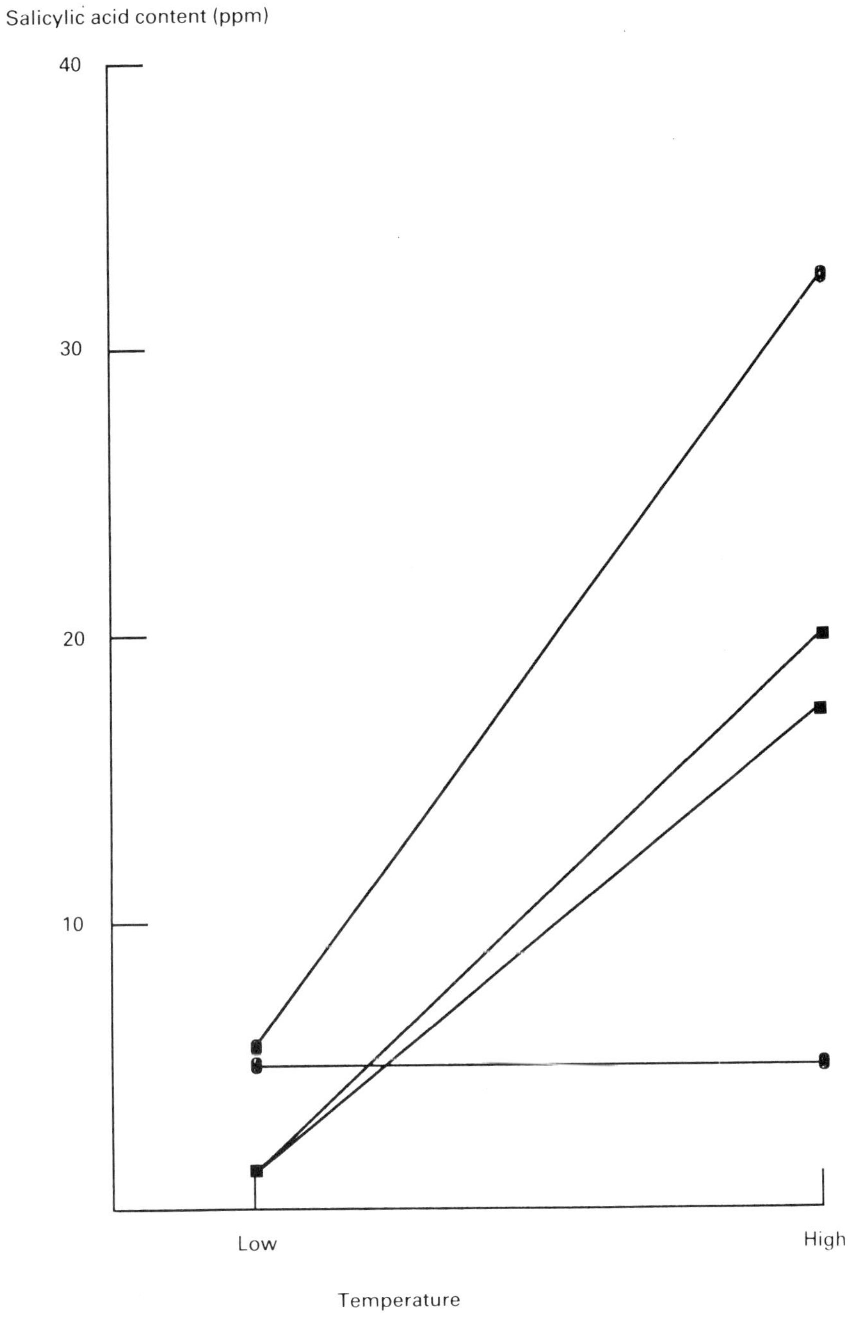

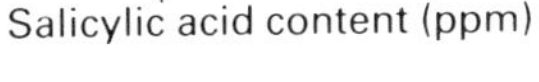

Fig. 4.3 — Salicylic acid content of aspirin tablets packed in glass (■) and PVC (●).

Table 4.5 — The effect on aspirin stability of temperature and humidity

Packaging	Effect of temperature	Effect of humidity	Interaction between temperature and humidity
PVC	13.15	13.95	13.15
Glass	16.35	1.25	1.35

FACTORIAL DESIGNS WITH THREE FACTORS

The previous discussion was limited to two experimental factors and a possible interaction between them. However, the principles of factorial design can be extended to situations where many more factors can be examined.

Consider the situation where three factors are suspected of having an influence on the outcome. The procedures involved are best shown by means of a worked example.

Lactose is a commonly used diluent for solid dosage forms. Though relatively inert, it can take part in the Maillard reaction to form brown pigments, which in turn cause discoloration of the dosage form. Among the factors which may affect the rate of the reaction and hence the degree of discoloration are temperature, humidity and the presence of a base, since the Maillard reaction is favoured by alkaline conditions.

Discoloration has been a particular problem with tablets containing spray-dried lactose, and Armstrong and Cartwright (1984) examined varieties of lactose, both spray and conventionally dried, to determine their propensity to develop a brown colour.

The following factors and levels were selected.

Factor *A*: absence or presence of a base (benzocaine).
Factor *B*: temperature; 25°C and 40°C.
Factor *C*: humidity; 50% and 75% RH.

As usual, low levels of a factor are represented by a − and high levels by a +. The experiments were set up as shown in Table 4.6, and after 2 months' storage in

Table 4.6 — Factorial design to show discoloration of lactose tablets

Experiment	Factor A (base)	Factor B (temperature)	Factor C (humidity)	Tablet colour
(1)	−	−	−	1.6
a	+	−	−	5.3
b	−	+	−	3.4
ab	+	+	−	6.6
c	−	−	+	2.6
ac	+	−	+	3.6
bc	−	+	+	3.0
abc	+	+	+	7.0

these conditions tablet colour was measured by reflectance meter, pure white being zero. The greater the degree of discoloration, the higher the number. The results given in the table are those for lactose monohydrate.

Thus, for example, experiment *ab* is carried out in the presence of benzocaine, the storage temperature is 40°C and the relative humidity is 50%.

Several points are worth making at this stage. Firstly note that the tablet colour must be expressed as a numerical value. Adjectival descriptions such as white, light brown, etc. cannot be used in designs of this type. Equally unacceptable are rank orders such as white=1, the next lightest tablet=2 etc.

Secondly note the standard order of the experiments in the table. The reason for adherence to this order will be apparent later.

Possible interactions must now be considered. In this case there are three two-way interactions (A with B, A with C, and B with C) and one three-way interaction (A with B and C). The signs of these interactions are determined by normal algebraic rules, and the overall design is shown in Table 4.7.

Table 4.7 — Signs to calculate main effects and interactions of a three-factor factorial design

Experiment	Factor			Interaction				Tablet colour
	A	B	C	AB	AC	BC	ABC	
(1)	−	−	−	+	+	+	−	1.6
a	+	−	−	−	−	+	+	5.3
b	−	+	−	−	+	−	+	3.4
ab	+	+	−	+	−	−	−	6.6
c	−	−	+	+	−	−	+	2.6
ac	+	−	+	−	+	−	−	3.6
bc	−	+	+	−	−	+	−	3.0
abc	+	+	+	+	+	+	+	7.0

Note that each column must have an equal number of + and − signs. This is a useful check that signs have been correctly allocated.

It is often useful to depict a factorial experiment of this type as a cube (Fig. 4.4). All experiments with a high level of factor A (a, ab, ac, abc) appear on the right-hand face of the cube and all with a low level of the same factor ((1), b, c, bc) on the left-hand face. Similarly, factor B is represented on the top and bottom faces of the cube, and factor C on the front and back faces.

The magnitudes of the main effects of the factors and the interactions can now be calculated. The method is exactly the same as before. Thus, for a factor A, the magnitude is the mean of all experiments with a high level of A minus the mean of all those with a low level.

Thus the magnitude of factor A is

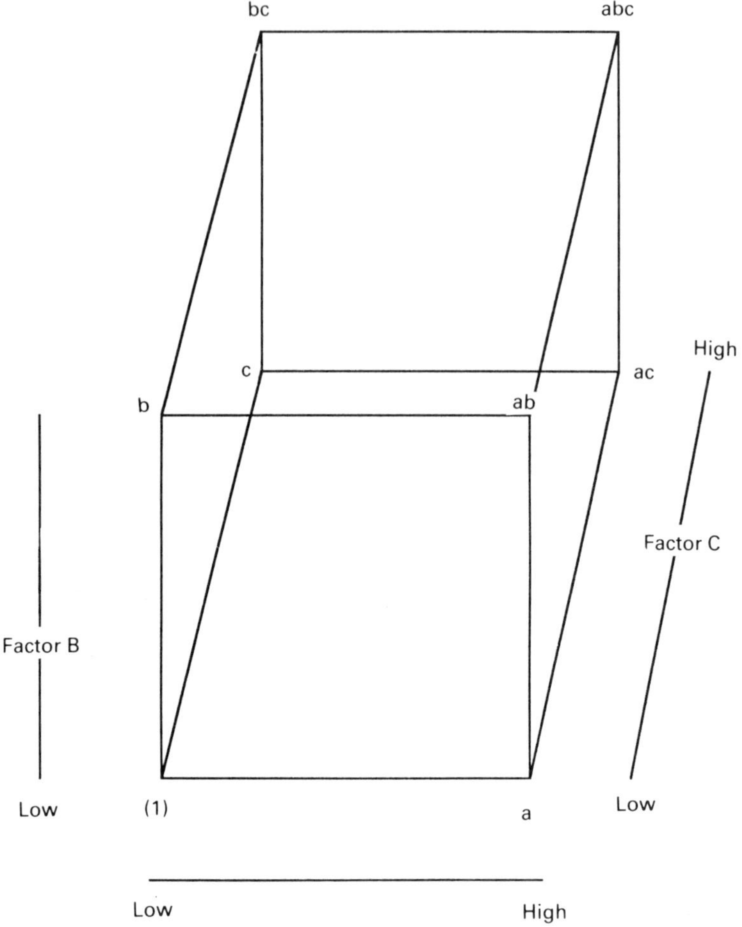

Fig. 4.4 — Representation of a three-factor, two-level experimental design.

$$1/4[a+ab+ac+abc]-[(1)+b+c+bc]$$
$$=1/4[(5.3+6.6+3.6+7.0)-(1.6+3.4+2.6+3.0)]$$
$$=2.975.$$

The complete set of values for main effects and interactions is given in Table 4.8.

Table 4.8 — Magnitudes of the main effects and interactions of the factors given in Table 4.7

Factor			Interaction			
A	B	C	AB	AC	BC	ABC
2.975	1.725	−0.175	0.625	−0.475	0.175	0.875

Thus the most important factor is the presence or absence of a base and the storage temperature. The environmental humidity is of less importance as are interactions between base and humidity and temperature and humidity. These conclusions however are based on a subjective assessment of the values shown in Table 4.8. Factorial design becomes an even more powerful technique when allied to analysis of variance, because then an objective assessment of the relative importance of the various factors and interactions can be obtained.

The most useful technique is that first described by Yates and is best demonstrated by using the same worked example as before.

The experimental data are first tabulated in standard order. Then the first two numbers (relating to experiments (1) and a) are added together and put into column 1 of Table 4.9a The next two are then added (experiments *b* and *ab*) and the result

Table 4.9a — Commencement of Yates's treatment for data from a factorially designed experiment

Experiment	Tablet colour	Column 1
(1)	1.6	6.9
a	5.3	10.0
b	3.4	6.2
ab	6.6	10.0
c	2.6	3.7
ac	3.6	3.2
bc	3.0	1.0
abc	7.0	4.0

put into the second row of column 1. Similarly, with the next two pairs (*c* and *ac*, *bc* and *abc*). Then the differences between adjacent pairs are calculated ($a-(1)$, $ab-b$, etc.) and these are placed into the fifth to the eighth rows of column 1.

At this stage, Table 4.9a appears as shown.

The process is then repeated using the numbers in column 1, and the results are placed into column 2. Thus the first number in column 2 is 16.9, obtained by adding together the first two rows in column 1, namely 6.9 and 10.0. The difference between these two numbers, 3.1, forms the fifth row of column 2. The identical process is repeated yet again on the numbers in column 2, the results being placed in column 3. The result is Table 4.9b.

Column 3 is now divided by 4, which is 2^{n-1}, where n is the number of factors examined (in this case 3). These results, the average effects, are put into column 4. Finally, the mean square is obtained by squaring the numbers in column 3 and

Table 4.9b — The second stage in Yates's treatment for data from factorially designed experiments

Experiment	Tablet colour	Column 1	Column 2	Column 3
(1)	1.6	6.9	16.9	—
a	5.3	10.0	16.2	11.9
b	3.4	6.2	6.9	6.9
ab	6.6	10.0	5.0	2.5
c	2.6	3.7	3.1	−0.7
ac	3.6	3.2	3.8	−1.9
bc	3.0	1.0	−0.5	0.7
abc	7.0	4.0	3.0	3.5

dividing by 2^n. Thus the mean square attributable to experiment *a* is $(11.9)^2/8=17.7$. The mean square is put into column 5, and the table now becomes Table 4.9c.

Table 4.9c — The final stage in Yates's treatment for factorial data

Experiment	Tablet colour	Column 1	Column 2	Column 3	Column 4	Column 5
(1)	1.6	6.9	16.9	—	—	—
a	5.3	10.0	16.2	11.9	2.975	17.70
b	3.4	6.2	6.9	6.9	1.725	5.95
ab	6.6	10.0	5.0	2.5	0.625	0.78
c	2.6	3.7	3.1	−0.7	−0.175	0.06
ac	3.6	3.2	3.8	−1.9	−0.475	0.45
bc	3.0	1.0	−0.5	0.7	0.175	0.06
abc	7.0	4.0	3.0	3.5	0.875	1.53

The importance of listing the experiments in standard order should now be apparent. Also it will be noted that the values in column 4 are those of the main effects and interactions first shown in Table 4.8.

The mean squares can now be placed in an analysis of variance table (Table 4.10). In any factorial of the form 2^n, each effect and interaction has one degree of freedom. It remains to calculate F, the ratio between the mean squares and the residual squares, also known as the error mean square. If the whole experiment has been replicated, then a number of observations will be available for each experiment and so an estimate of the experimental error can be made. This is undoubtedly the favoured approach. However, replication may lead to an acceptably high number of experimental runs. In these circumstances, the usual approach is to assume that some

Table 4.10 — Analysis of variance table following Yates's treatment

Factor or interaction	Experiment	Degrees of freedom	Mean square	F
Base	*a*	1	17.70	295
Temperature	*b*	1	5.95	99
Humidity	*c*	1	0.78	13
Base × temperature	*ab*	1	0.06	—
Base × humidity	*ac*	1	0.45	7
Temperature × humidity	*bc*	1	0.06	—
Base × temperature × humidity	*abc*	1	1.53	25

interactions have a negligible effect, and so experimental runs containing these can be combined to give the experimental error. Alternatively, results which give very low values in the mean squares column may be combined for this purpose. Naturally incorrect assumptions can be made, and factors and interactions which are truly significant can be assumed to be zero. A knowledge of the experimental system being studied and the use of common sense will help select those interactions which are likely to be of least significance. Adopting this approach to the current data, the mean squares relating to experiments *c* and *bc* are distinctly lower than the others. These can therefore be combined to give a mean of 0.06 as the experimental error of the system, and F is calculated by dividing the other mean squares by this number. Table 4.10 gives the complete analysis of variance table.

The significance of the values of F are assessed by comparing them with tabulated values. The numerator has one degree of freedom and the denominator has two. Therefore for $p<0.05$, F should exceed 18.5. For $P<0.01$, F should be greater than 98.5. Thus the presence of base is clearly the most important factor.

Calculation of the main effects, interactions and the Yates technique for analysis of variance is facilitated by use of a computer. A program in BASIC for this purpose is given in Appendix 1.7.

FACTORIAL DESIGNS WITH MORE THAN THREE FACTORS

One of the key features of a factorially designed experiment is that all factors other than those under investigation should be kept constant. A corollary of this is that all experiments should be carried out at the same time by the same personnel using identical equipment, but this may not always be possible.

Consider the following hypothetical example. A dissolution test is being evaluated. Concern has been expressed that the concentration of a dissolved drug is apparently influenced by the position in the apparatus from which the sample is taken. It has been suggested that both the horizontal and vertical distance from the dissolution cell are factors which may have influence. A two-factor, two-level experiment is designed (Fig. 4.5).

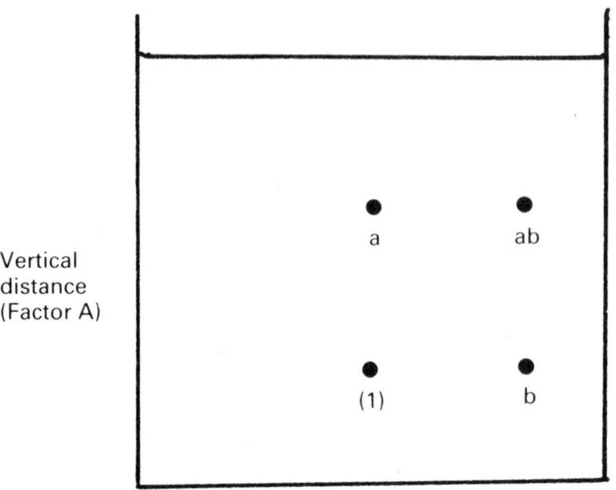

Fig. 4.5 — Two-factor, two-level design for dissolution.

Vertical distance is designated factor A and horizontal distance factor B. Samples of the dissolution fluid are to be extracted by sampling probes. Unfortunately, due to the design of the sampling apparatus, it is impossible to extract more than two samples at any one time. However, if the samples are extracted as two pairs, then dissolution will have occurred between extraction of the first pair and that of the second pair.

It is possible to arrange the two pairs of samples in three ways. These are

(1) and *a*, *b* and *ab*
(1) and *b*, *a* and *ab*
(1) and *ab*, *a* and *b*.

Thus, in an attempt to evaluate horizontal displacement, samples (1) and *a* are withdrawn, followed by *b* and *ab*. The second pair may differ from the first for two reasons, firstly because horizontal displacement does play a role and secondly because time has elapsed between taking the two pairs of samples. This confusion is known as confounding.

By the same argument, if the first pair of samples is (1) and *b*, the vertical displacement will be confounded with sampling time, since all the high level readings are in one pair and the low level readings are in the other. The third arrangement is for the first pair to be (1) and *ab*, and the second *a* and *b*. In this neither main effect is confounded but the interaction is. As a general rule, main effects should not be confounded, and if confounding is unavoidable, it is better to confound an interaction, which of course may not be present in any case.

In the above example, confounding is unavoidable because of constraints imposed by the equipment. However, confounding can often be used to advantage to reduce the number of experiments which must be carried out. Four experiments are required for a full two-factor, two-level design and eight experiments are required for a similar three-factor design, i.e. if there are N factors at two levels, then 2^N experiments are needed for a full design. Thus the number of experiments can grow rapidly, and the consequent high consumption of time and materials may nullify the advantages of the factorial approach. However, if it can be decided at the outset that certain features of the design can safely be confounded, it is possible to run fewer combinations of factors than is theoretically necessary. Even so it must be clearly understood that a price must be paid, and a full evaluation of all factors and all interactions cannot be made if a confounded design is used.

The use of confounded systems is particularly useful when the number of factors under investigation is large.

A number of methods of drawing up blocks of experiments have been devised. Their derivation is beyond the scope of this publication, but some examples are shown below.

An example of a three-factor, two-level factorial in two blocks is shown in Table 4.11. Consideration of the table shows that block 1 contains all those combinations in

Table 4.11 — Three-factor two-level factorial design in two blocks

Block 1	Block 2
(1)	*a*
ab	*b*
ac	*c*
bc	*abc*

which *abc* is positive and block 2 all those when the same interaction is negative. Thus the three-way interaction is confounded with the blocks. It will be recalled that in earlier examples, the three-way interaction was often used as the 'error' term in designs of this type, i.e. it was assumed to have negligible significance. Hence its confounding cannot be regarded as a major loss.

A two-level four-factor design is shown in Table 4.12. Interaction ABCD is confounded with the blocks, with 1 degree of freedom, and the four three-way interactions could be pooled with 4 degrees of freedom as the error term. A similar design, but now in four blocks, is given in Table 4.13. Here the blocks are confounded with ABCD, BCD and AD. The blocks and their interactions account for 3 degrees of freedom and the error term could be ABD, ACD and ABCD, also with 3 degrees of freedom.

Reduction in the number of experiments can also be achieved by choosing a fractional factorial design. Thus if there are four factors, A, B, C and D, but only

Table 4.12 — Four-factor two-level factorial design in two blocks

Block 1	Block 2
(1)	*a*
ab	*b*
bc	*abc*
ac	*c*
abcd	*bcd*
cd	*acd*
ad	*d*
bd	*abd*

Table 4.13 — Four-factor, two-level factorial design in four blocks

Block 1	Block 2	Block 3	Block 4
(1)	*a*	*b*	*ab*
bc	*abc*	*c*	*ac*
acd	*cd*	*abcd*	*bcd*
abd	*bd*	*ad*	*d*

eight experiments can be carried out from the possible 16 combinations, a suitable fractional design is shown in Table 4.14.

Table 4.14 — Four-factor, two-level factorial fractional design

Factor				Treatment
A	B	C	D	combination
−	−	−	−	(1)
+	−	−	+	*ad*
−	+	−	+	*bd*
+	+	−	−	*ab*
−	−	+	+	*cd*
+	−	+	−	*ac*
−	+	+	−	*bc*
+	+	+	+	*abcd*

Because of the reduction in the number of experiments, considerable confounding has occurred. Thus:

Main effect A is confounded with BCD
Main effect B is confounded with ACD
Main effect C is confounded with ABD
Main effect D is confounded with ABC
Interaction AB is confounded with interaction CD
Interaction AC is confounded with interaction BD
Interaction BC is confounded with interaction AD

Note that there are three pairs of two-factor interactions. The results of such a design can be analysed by the Yates method.

Further fractional combinations are given in Table 4.15a and 4.15b.

Table 4.15a — Six-factor, two level factorial design in eight experiments

Factor						Treament
A	B	C	D	E	F	combination
−	−	−	+	+	+	*def*
+	−	−	−	−	+	*af*
−	+	−	−	+	−	*be*
+	+	−	+	−	−	*abd*
−	−	+	+	−	−	*cd*
+	−	+	−	+	−	*ace*
−	+	+	−	−	+	*bcf*
+	+	+	+	+	+	*abcdef*

Table 4.15b — Seven-factor, two level factorial in eight experiments

Factor							Treatment
A	B	C	D	E	F	G	combination
−	−	−	+	+	+	−	*def*
+	−	−	−	−	+	+	*afg*
−	+	−	−	+	−	+	*beg*
+	+	−	+	−	−	−	*abd*
−	−	+	+	−	−	+	*cdg*
+	−	+	−	+	−	−	*ace*
−	+	+	−	−	+	−	*bcf*
+	+	+	+	+	+	+	*abcdefg*

Techniques involving even more factors are available, mainly devised by Plackett and Burman (1946). They prepared two-level factorial designs for studying $(N-1)$ variables in N experiments, where N is a multiple of four. If N is a power of 2, the designs are identical to those already discussed.

Table 4.16 gives the rows of + and − signs used to construct Plackett–Burman

Table 4.16 — Plus and minus signs for Plackett–Burman designs

N=12++−+++−−−+−
N=20++−−++−+−+−+−−−−++−
N=24+++++−+−++−−++−−+−+−−−−

designs for N=12, 20 and 24. The complete designs are obtained by writing the relevant row as a column. The next column is generated by moving the elements down one row, and placing the last element in the first position. Subsequent columns are prepared in the same way. Finally the design is completed by adding a row of minus signs. A Plackett–Burman design for studying eleven factors in twelve experiments is given in Table 4.17. If the number of factors is less than 11 but greater

Table 4.17 — Plackett–Burman design for eleven experiments

					Factor					
A	B	C	D	E	F	G	H	I	J	K
+	−	+	−	−	−	+	+	+	−	+
+	+	−	+	−	−	−	+	+	+	−
−	+	+	−	+	−	−	−	+	+	+
+	−	+	+	−	+	−	−	−	+	+
+	+	−	+	+	−	+	−	−	−	+
+	+	+	−	+	+	−	+	−		−
−	+	+	+	−	+	+	−	+	−	−
−	−	+	+	+	−	+	+	−	+	−
−	−	−	+	+	+	−	+	+	−	+
+	−	−	−	+	+	+	−	+	+	−
−	+	−	−	−	+	+	+	−	+	+
−	−	−	−	−	−	−	−	−	−	−

than 8, then 12 experiments must still be carried out. However, replicates can be incorporated, providing the error term in subsequent analysis. Considerable confounding occurs in all Plackett–Burman designs.

Occasionally finding that an interaction is real can have a beneficial result. If for example an interaction occurs between the temperature of a reaction and the quantity of catalyst present, then yield is highest either with a low temperature and large amount of catalyst or high temperature and small amounts of catalyst. Hence the most economical solution depends on the relative costs of energy and catalyst.

FACTORIALS WITH THREE LEVELS

The applications of factorial analysis so far described deal with only two levels of a particular factor, e.g. high and low, or presence and absence. This implies that there is a straight-line relationship between the magnitude of the factor and its effect. If this is known not to be the case, then it may be necessary to use more than two levels. However, this approach should be treated with caution, as it markedly increases the number of experiments. If a three-level design is used, these levels can be designated as low, intermediate and high. Obviously the previously used notation, whereby a particular experiment is described by a lower case letter when that factor is at a high level, cannot be used in this case. The levels are therefore designated numerically, viz. 0 (low), 1 (intermediate) and 2 (high). Thus any particular experiment is designated by a two-digit number. For example, 00 has both factors at their lowest level and 12 has the first factor at the intermediate level and the second at its highest level. Experiments with more than three levels can also be designated using this numerical system. There is no reason why the numerical system cannot be used with two-level designs, but the vast majority of literature on factorial analysis uses an alphabetical notation for two-level studies.

Factors are usually designated by capital letters as before, and also as before it is often useful to envisage the experimental design in diagrammatic form. Thus Fig. 4.6 represents a two-factor, three-level design.

The use of three levels implies non-linear relationships between the two factors and the response. These can be expressed in the form of equations (4.1) and (4.2), which contain both linear terms (A and B to the power 1) and quadratic terms (A and B to the power 2). These are usually designated A_L and B_L, and A_Q and B_Q respectively.

$$\text{Response} = a + bA + cA^2 \tag{4.1}$$

$$\text{Response} = a + bB + cB^2 \tag{4.2}$$

Combination of these equations gives interaction terms such as $A_L \times B_L$, $A_Q \times B_L$ etc., and a full analysis includes determination of these as well as the main effects.

As before, this will be demonstrated by means of a worked example.

A pressurized inhalation device delivers droplets to the lung with a wide spectrum of sizes. Only some of these droplets can be deposited in the required part of the respiratory tract, and these are known as the respirable fraction. The magnitude of this fraction is governed by a number of factors. Two of these are the concentration of surfactant in the system and the concentration of water. A third important factor is the design of the valve on the pack, and this will be introduced into the discussion

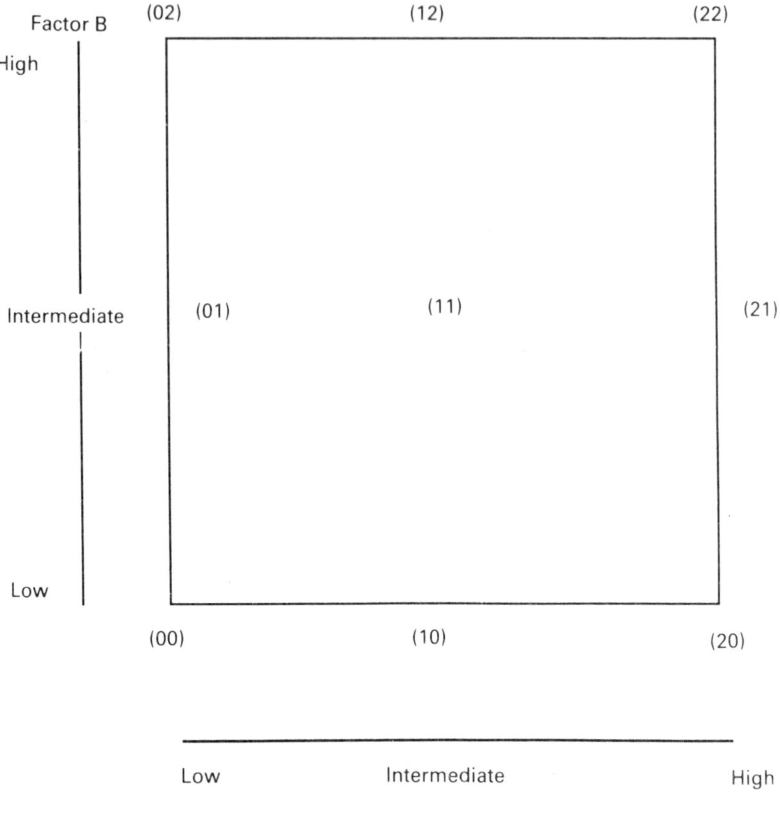

Fig. 4.6 — Representation of a two-factor, three-level experimental design.

later. However, for the present example, it will be assumed that the same design of valve is used throughout.

Let surfactant concentration be designated factor A, and three levels chosen (0.5, 1.0 and 1.5%). These are designated levels 0, 1 and 2 respectively. Water concentration is designated factor B, and the levels are 1.4, 2.8 and 4.2%. Thus the design of the experiment is shown in Table 4.18. The treatment combinations are shown in the top left-hand corner of each cell and the corresponding values for the respirable fraction are given in the bottom right-hand corner of the same cell. Column and row totals are also given.

The procedure is as follows. Analysis of the data gives the sums of squares etc. Thus:

$$SS_{\text{total}} = 52.5^2 + 50.0^2 + \ldots 24.1^2 - \frac{397.5^2}{9}$$

Table 4.18 — Two-factor, three-level factorial design. Relationship between respirable fraction, surfactant concentration and water concentration

Factor B	Factor A						Total
		0		1		2	
	00		10		20		
0		52.5		50.0		39.1	141.6
	01		11		21		
1		53.2		47.9		33.1	134.2
	02		12		22		
2		56.3		41.3		24.1	121.7
Total		162.0		139.2		96.3	397.5

$$= 18461.51 - 17556.25$$

$$= 905.26$$

$$SS_A = \frac{162.0^2 + 139.2^2 + 96.3^2}{3} - \frac{397.5^2}{9}$$

$$= 741.86$$

$$SS_B = \frac{141.6^2 + 134.2^2 + 121.7^2}{3} - \frac{397.5^2}{9}$$

$$= 67.44$$

$$SS_{error} = 905.26 - (741.86 + 67.44)$$

$$= 95.96$$

The analysis of variance table (Table 4.19) can now be constructed.

Table 4.19 — Analysis of variance table for data given in Table 4.18

Source	df	Sum of squares	Mean square
A	2	741.86	370.93
B	2	67.44	33.72
AB	4	95.96	23.99
Total	8	905.26	—

Both A and B have linear and quadratic terms, and so analysis can now be taken further. The responses are multiplied by the coefficients given in Table 4.20. The

Table 4.20 — Coefficient for a two-factor, three level factorial design

	Treatment combination									Sum of coefficients squared
Factor	00	01	02	10	11	12	20	21	22	
A_L	−1	−1	−1	0	0	0	+1	+1	+1	6
A_Q	+1	+1	+1	−2	−2	−2	+1	+1	+1	18
B_L	−1	0	+1	−1	0	+1	−1	0	+1	6
B_Q	+1	−2	+1	+1	−2	+1	+1	−2	+1	18
A_LB_L	+1	0	−1	0	0	0	−1	0	+1	6
A_LB_Q	−1	+2	−1	0	0	0	+1	−2	+1	12
A_QB_L	−1	0	+1	+2	0	−2	−1	0	+1	12
A_QB_Q	+1	−2	+1	−2	+4	−2	+1	−2	+1	36

derivation of these coefficients is beyond the scope of this text, but appropriate references for further study are given in the bibliography.

Consider the first row of Table 4.20. This refers to the linear effect of factor A, and compares the three lowest levels of A (00, 01, 02) with the three highest levels of A (20, 21, 22), taken across all levels of B. The coefficients of the interaction terms are obtained by multiplying together those of the main effects. These coefficients are now applied to the responses given in Table 4.18.

Thus:

$$A_L=(52.5\times-1)+(53.2\times-1)+(56.3\times-1)+(50.0\times0)+(47.9\times0)+(41.3\times0)+(39.1\times+1)+(33.1\times+1)+(24.1\times+1)$$
$$=-65.7.$$

The corresponding sum of squares is $(65.7)^2/6=719.4$.

By identical methods, the multiples of the other responses and coefficients, and sums of squares, are calculated, and are summarized in Table 4.21, which also contains analysis of variance data.

The absence of replication precludes a proper calculation of the underlying error of the system. However, if the smallest mean squares (B_Q, A_LB_Q, A_Q,B_L, A_Q B_Q) are averaged, that can form the denominator of the F ratio. These F values are given in Table 4.21. Thus A_L, A_Q, B_L and A_LB_L are all significant at $P=0.05$ and all except A_Q at $P=0.01$. From this it can be inferred that since all quadratic terms of the main factors and interactions are of low significance, a reasonably linear relationship links both factors and the response, though interaction between the main effects is significant.

Table 4.21 — Sums of squares and ANOVA for data from Table 4.18

Source	Response× coefficient	Sum of squares		df		Mean squares		F
A		741.8		2		370.9		
A_L	−65.7		719.4		1		719.4	319
A_Q	−20.1		22.4		1		22.4	10
B		67.4		2		33.7		
B_L	−19.9		66.0		1		66.0	29
B_Q	−5.1		1.4		1		1.4	
AB		96.0		4		24.0		
A_LB_L	−18.8		88.4		1		88.4	39
A_LB_Q	−5.4		2.4		1		2.4	
A_QB_L	6.2		3.2		1		3.2	
A_QB_Q	8.4		2.0		1		2.0	

Blocked designs for three-level factorials are also available. The usual procedure is to arrange the experiments in blocks which are multiples of three. Thus a three-level, three-factor design is arranged in three blocks of nine experiments as in Table 4.22. In this example, the numerical notation is used. All main effects (A, B and C)

Table 4.22 — Three-factor, three-level factorial design in three blocks

Block 1	Block 2	Block 3
000	100	200
011	111	211
110	210	010
121	221	021
102	202	002
212	012	112
220	020	120
022	122	222
201	001	101

and all two-way interactions can be isolated. Examples like this can be evaluated using the Yates method as described earlier.

THREE-FACTOR, THREE-LEVEL FACTORIAL DESIGNS

The next stage in complexity is the three-factor design where each factor is studied at three levels. The experimental layout and notation are shown in Fig. 4.7. As there

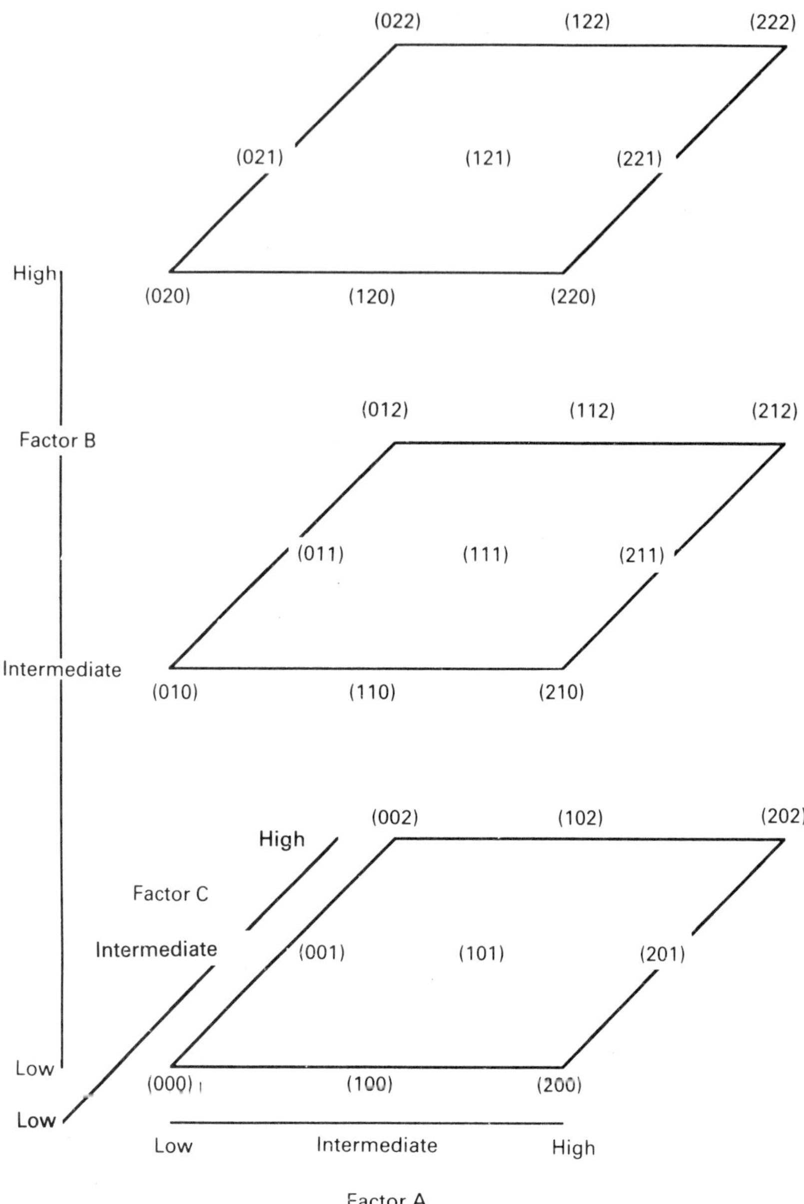

Fig. 4.7 — Representation of a three-factor, three-level experimental design.

are 27 possible combinations, there are 26 degrees of freedom. Each main effect has two degrees of freedom, the two factor interactions have four degrees of freedom each and the three-factor interaction has eight. If the factors are quantitative and equally spaced, the main effects can be partitioned into linear and quadratic components as before, as can the interaction terms. The eight possible combinations which can be derived from the three-way interaction ($A_L B_L C_L$, $A_Q B_L C_L$, etc.) are often difficult to explain on a practical basis, and so the ABC interaction often serves as the 'error' by which the main effects and two-way interactions are tested.

The pressurized pack example used earlier can usefully be extended into 3^3 factorial. The third factor to be introduced is the design of the valve. This can be qualitative, e.g. overall valve design, or quantitative, e.g. valve aperture size.

Thus surfactant concentration is factor A, water concentration is factor B and valve design is factor C.

The experimental design and data are shown in Table 4.23, subtotals and the grand total also being given.

Table 4.23 — Three-factor three-level factorial design. Data are respirable fraction (%)

	Factor A (surfactant)									
	0			1			2			
Factor C	Factor B (Water)									
(valve)	0	1	2	0	1	2	0	1	2	Total
0	52.5	53.2	56.3	50.0	47.9	41.3	39.1	33.1	24.1	397.5
1	46.2	53.8	33.5	52.9	43.4	19.1	47.0	32.7	18.5	347.1
2	40.3	38.6	29.4	39.7	34.2	21.9	40.2	30.5	15.0	289.8
	139.0	145.6	119.2	142.6	125.5	82.3	126.3	96.3	57.6	1034.4

The first stage in any analysis of variance is to calculate the correction term. This equals the square of the sum of all the terms, divided by the number of terms.

Thus the correction term$=(1034.4)^2/27$

$=39629.0$

The next stage is to calculate the sums of squares of the main effects A, B and C.

Thus for main effect A:

The sum of squares of all results when factor A is at level 0

$=(52.5+53.2+56.3+46.2+53.8+33.5+40.3+38.6+29.4)^2/9$

$=(403.8)^2/9$

$=18117.2$

Similarly, when factor A is at level 1, the sum of squares

$=(50.0+47.9+\ldots+21.9)^2/9$

$=350.4^2/9$

$=13642.2$

Similarly, when factor A is at level 2, the sum of squares is

$=(39.1+33.1+\ldots+15.0)^2/9$
$=280.2^2/9$
$=8723.6.$

Adding these three terms together and subtracting the correction term gives

$40483.0-39629.0=854.0$

Thus the sum of squares of factor $A=854.0$.
By identical methods the main effects of factors B and C can be calculated. Thus for factor B:

The sum of squares when B is at level 0
$=(407.9)^2/9$
$=18486.9$

The sum of squares when B is at level 1

$=(367.4)^2/9$
$=14998.1$

The sum of squares when B is at level 2

$=(259.1)^2/9$
$=7459.2$

Totalling these three terms and subtracting the correction term gives:

$18486.9+14998.1+7459.2-39629.0$
$=1315.2$

Similarly, the sum of squares of the main factor C is 645.4.

The next stage is to calculate the three two-factor interactions AB, AC and BC.

For the AB interaction, changes in C are ignored. For example, the results of experiments $A_0B_0C_0$, $A_0B_0C_1$ and $A_0B_0C_2$ are added together and the sum squared. Since there are three terms, the sum of squares is divided by three, and the correction term subtracted. From this is then subtracted the sums of squares of the main effects A and B. What remains is the sum of the squares of the interaction AB.

Expressing this in numerical terms:

The sum of squares of the AB interaction
$=(52.5+46.2+40.3)^2/3$
$+(53.2+53.8+38.6)^2/3$

$+\ldots(24.1+18.5+15.0)^2/3$
-39629.0
$-(854.0+1315.2)$
$=245.3$

Similarly, to calculate the AC interaction changes in factor B are ignored. Thus results from experiments $A_0B_0C_0$, $A_0\,B_1C_0$ and $A_0B_2C_0$ are grouped together.

Thus the sum of squares for the AC interaction

$=(52.5+53.2+56.3)^2/3$
$+(46.2+53.8+33.5)^2/3$
$+\ldots(40.2+30.5+15.0)^2/3$
-39629.0
$-(854.0+645.4)$
$=180.9$

For the BC interaction, the sum of the squares

$=297.1.$

The next stage is to calculate the sum of squares of the three-way interaction ABC.

This is done by calculating the sum of the squares of all the terms and subtracting those of the main effects and the two-way interactions. Thus

Sum of squares of the ABC interaction is
$(52.5^2+46.2^2\ldots+15.0^2)$
-39629.0 (the correction term)
$-(854.0+1315.2+645.4)$ (the three main effects)
$-(245.3+180.9+297.1)$ (the two-way interactions)
$=88.1.$

The ANOVA Table can now be constructed (Table 4.24).

Table 4.24 — ANOVA table for data presented in Table 4.23

Source	DF	Sum of squares	Mean square	*F*
A	2	854.0	427.0	38.8
B	2	1315.2	657.6	59.8
C	2	645.4	322.7	29.3
AB	4	245.3	61.3	5.6
AC	4	180.9	45.2	4.1
BC	4	297.1	74.3	6.8
ABC	8	88.1	11.0	—
Total	26	3626.1	—	—

In the absence of replication of individual data points, it is useful to use the *ABC* interaction with its eight degrees of freedom as the error term. Dividing all other mean squares by the mean square of the *ABC* interaction gives the value of *F* shown in the right-hand column of Table 4.24.

All main effects are significant at the 1% level of significance, but none of the interactions has significance even at the 5% level.

GENERAL COMMENTS ON FACTORIAL DESIGN

Like any other experimental technique, factorial designs repay thought and planning. Two crucial decisions to be taken *a priori* are the factors to be studied and the levels of these factors, and both of these depend on the objectives of the exercise. Factors not relevant to the experiment but which may affect the magnitude of the results should, as far as possible, be controlled. Such factors might be, for example, different equipment, different personnel or even different locations. Furthermore, environmental factors should also be controlled if it is suspected that they may affect results, or else they should be included in the design as an additional variable.

The more external factors that can be controlled, the lower will be the residual variation and hence the more valid any analysis of variance.

The choice of levels and their number is also a crucial factor. The increased complexity of a design if more than two levels of a particular variable are to be investigated has already been discussed. Nonetheless, it must be remembered that selection of only two levels implies a linear relationship between the magnitude of the factor and the magnitude of its effect over the range selected. Here again, judgment and common sense are vital. If the factor is qualitative, then there is usually no difficulty, i.e. the factor is either 'present' or 'absent'. Even here however with factors which are 'present', the level at which they are present can be vital. It is useless including so little of the factor that there is insufficient to have a detectable effect. With quantitative factors, a useful guide is to take the extremes of the useful range and divide it into quarters. The one quarter and the three-quarter values are then taken as the levels. For example, if the likely range of temperature is 20°C to 100°C, then suitable levels might be 40°C and 80°C.

When used properly, factorial design is a powerful tool. Maximum use is made of all the data, since, as is shown in the worked examples, all the data are used in the calculation of main effects and interactions. Factorial designs are also orthogonal, in that the estimated effects and interactions are independent of other factors in the experiment. However, confounded or partial designs cause loss of independence and, in such cases, interactions can be difficult to detect.

REFERENCES

Armstrong, N. A. and Cartwright, R. G. (1984) The discoloration on storage of tablets containing spray-dried lactose, *J. Pharm. Pharmacol.* **36** 5P.

Plackett, R. L. and Burman, J. P. (1946) The design of optimum multifactorial experiments, *Biometrika* **33** 305–325.

FURTHER READING

Bolton, S. (1983) Factorial designs in stability studies, *J. Pharm. Sci.* **72** 362–366.

Jones, K. (1986) *Optimisation of Experimental Data*, International Laboratory, November 1986, pp. 32–45.

Malinowski, H. J. and Smith, W. E. (1975) Use of factorial design to evaluate granulation prepared by spheronisation, *J. Pharm. Sci.* **64** 1688–1692.

Montgomery, D. C. (1976) *Design and Analysis of Experiments*, Wiley.

Plazier-Vercammen, J. A. and De Neve, R. E. (1980) Evaluation of complex formation by factorial analysis, *J. Pharm. Sci.* **69** 1403–1408.

Sanderson, I. M., Kennerley, J. W. and Parr, G. D. (1984) An evaluation of the relative importance of formulation and process variables using factorial design, *J. Pharm. Pharmacol.* **36** 789–795.

5

Correlation and regression

LINEAR REGRESSION

Before the development of computers, rectilinear relationships were detected by plotting a graph of one variable against another and observing if the points could reasonably be considered to follow a straight line. If they did, the best straight line was judged subjectively, and drawn through the points with a ruler. When required, slope and intercept were measured by counting the squares on the graph paper, and intermediate values were determined by interpolation. Proof of rectilinearity was demonstrated in reports and research papers by showing the graph. A method for calculating the best equation relating the points, known as regression or least squares analysis, was known, but was protracted, particularly when there were a large number of points, and when many relationships had to be examined. However, the calculations can be done quickly with computers, so that nowadays the procedure is programmed into most computers, and even into the more sophisticated pocket calculators.

Table 5.1, which gives the viscosities of a range of glycerol–water mixtures at 23°C (Gebre-Mariam 1988), can be used to demonstrate the process of linear regression analysis. The plot of the data is shown in Fig. 5.1. The points follow a slightly curved course, which is sufficiently shallow to be assumed to be rectilinear, so that a straight line can be drawn between them, as shown. The procedure offers a reasonable interpretation of the information, but if the points were more scattered

Table 5.1 — Viscosities of glycerol–water mixtures at 23°C

% w/w glycerol	12.3	18.5	24.6	30.8	36.9
Viscosity $Nm^{-2}s \times 10^3$	4.83	6.32	7.50	9.66	11.9

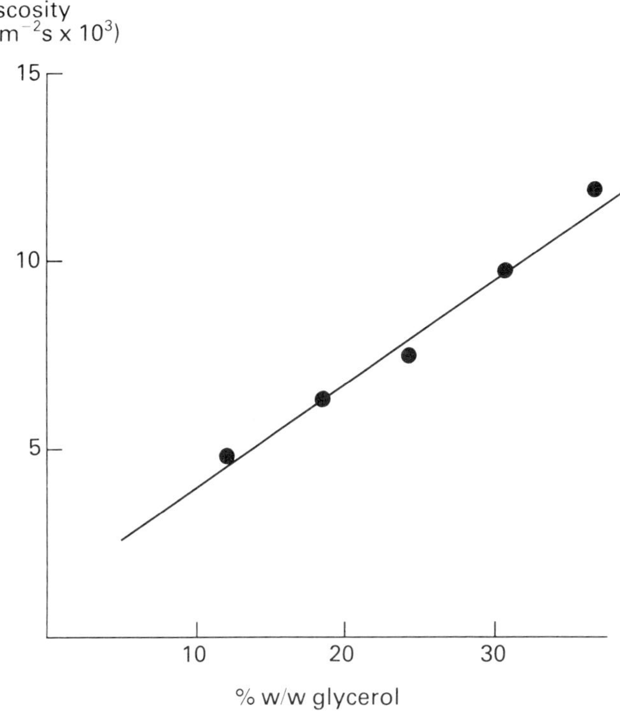

Fig. 5.1 — Viscosities of glycerol–water mixtures at 23°C.

the orientation of the line would become subjective, and dependent on the judgement of the observer.

The percentages of glycerol in the mixture are quantities with negligible errors, but viscosity is a quantity liable to random error. By convention, the reliable property takes the horizontal (x) axis, and the random quantity takes the vertical (y) axis. The equation of the best-fitting straight line, having the form shown in equation (5.1), is called the regression line, and can be calculated by least squares or regression analysis. b_1 represents the slope of

$$y = b_0 + b_1 x \tag{5.1}$$

the line, and b_0 the intercept. The regression line is that for which the sum of the vertical distances between the experimental points and the line is less than the equivalent sum obtained with any other straight line. If the vertical distances are expressed as the observed result (y_{obs}) minus the result predicted by the line (y_{pred}), some will be positive and some will be negative, and will therefore cancel each other. The problem is overcome by squaring the distances, so that all will become positive before summation, hence the term, least squares analysis. The slope (b_1) is of the regression line and is calculated from equation (5.2). b_1 is the regression coefficient.

$$b_1 = \frac{S(xy) - S(x)S(y)/N}{S(x^2) - S^2(x)/N} \tag{5.2}$$

$S(xy)$ is the sum of the products of each value of x and the corresponding value of y, so that for the present information,

$$S(xy) = (12.3 \times 4.83) + (18.5 \times 6.32) + \ldots (36.9 \times 11.9)$$
$$= 1097.467$$

$S(x)S(y)/N$ is the sum of all the xs, multiplied by the sum of all the ys, divided by the number of pairs of x and y. Thus, since $S(x) = 123.1$, $S(y) = 40.21$ and $N = 5$.

$$S(x)S(y)/N = \frac{123.1 \times 41.212}{5} = 989.97$$

$S(x^2)$ is the sum of the squares of each value of x, i.e.

$$S(x^2) = 12.3^2 + 18.5^2 + \ldots + 36.9^2 = 3408.95$$

$S^2(x)/N$ is the sum of all the ys squared, divided by the number of pairs of x and y, i.e.

$$S^2(x)/N = \frac{123.1^2}{5} = 3030.72$$

Collecting all these together,

$$b_1 = \frac{1097.47 - 989.97}{3408.95 - 3030.72} = 0.284$$

Substituting 0.284 for b_1, together with the mean value of x (= 24.6) and the mean value of y (= 8.042) into equation (5.1), yields $b_0 = 1.045$, giving the regression data for the equation (5.3).

$$y = 0.284x + 1.05 \tag{5.3}$$

A regression equation is a convenient alternative to a graph for expressing a relationship between variables, and is easily used for interpolation by substituting x with the concentration of the solution whose viscosity is to be predicted. However, used on its own, it does not give as much information as a graph, because the appearance of a graph indicates how many results were used, how scattered the results are about the line, and how representative of the results the line is. Additional statistical parameters should therefore be given with the basic equation, to answer these questions. A typical format is as follows,

$$\begin{array}{lllll} & & N & r & s \\ y = 0.284\ (0.013)x + 1.05\ (0.562) & & 5 & 0.991 & 0.379 \\ \quad 21.2 \qquad\qquad 1.87 & & & & \\ F_{1,3} = 174.2 \qquad \alpha,0.01 = 34.1 & & & & \end{array} \tag{5.4}$$

N represents the number of pairs of variables used to calculate the regression equation. Obviously, the greater the value of N, the more reliable the equation as a means of predicting new information.

r is the correlation coefficient. It is usually quoted as a number which varies from zero to 1. The higher the number, the greater the likelihood that x and y are related. What constitutes a satisfactory correlation coefficient depends on the value of N; the greater N is, the lower the acceptable correlation coefficient, and on the purpose for which the results are to be used. For a linear regression between two physical properties involving five pairs of results, as in the above example, a correlation coefficient in excess of 0.990 would be sought, but if looking for a relationship between percentage glycerol and a biological response, 0.950 could be a suitable target. In some procedures, for example cluster analysis, considerably lower correlation coefficients can be considered important.

A mistaken idea which is frequently assumed is that any collection of points which precisely coincides with a straight line will have a correlation coefficient of unity. This is not the case; the correlation coefficient assesses if the two sets of variables are related. This can be illustrated by considering the densities of water between 4° and 12°C, shown in Table 5.2. If both sets of variables are rounded off to two significant

Table 5.2 — Densities of water at various temperatures

Temperature (°C)	4.00	6.00	8.00	10.00	12.00
Density (kg m^{-3})	1000.00	999.97	999.88	999.73	999.53

figures, and plotted against each other, the graph in Fig. 5.2(a) will be obtained. The densities do not appear to change, despite a threefold increase in temperature, and the graph takes the form of a perfect rectilinear plot, but indicates that the density of water is independent of temperature over this range. This is confirmed by the correlation coefficient (r), which is expressed by equation (5.5). s_x and s_y are the standard deviations of x and y.

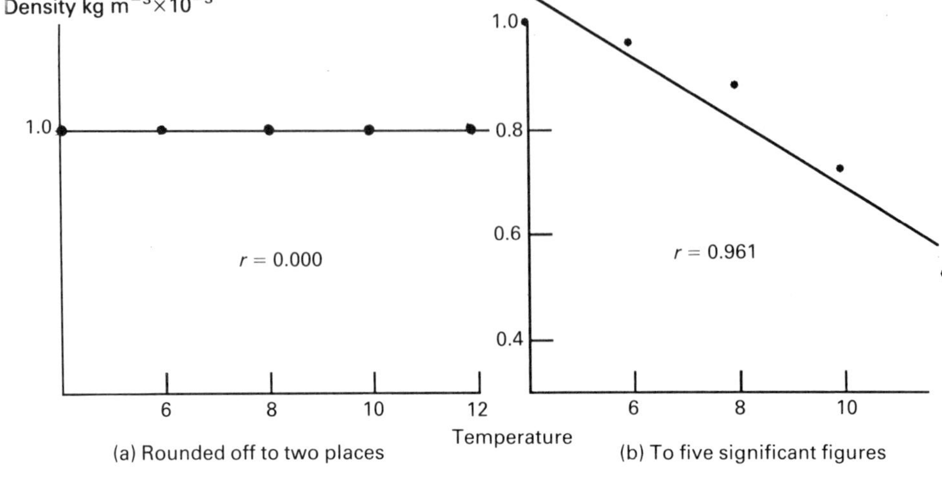

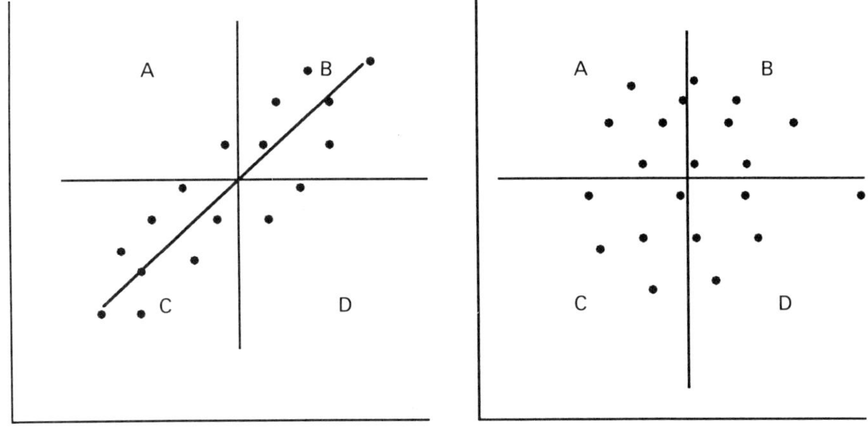

Fig. 5.2 — Regression. (a) and (b), densities of water at various temperatures.

$$r = \frac{1}{N} S \frac{(x - \bar{x})(y - \bar{y})}{s_x s_y} \tag{5.5}$$

The mean density ($\bar{y}$) is equal to 1.0×10^3 as also are all the values of y, so that $(y - \bar{y}) = 0$, and the whole expression for r reduces to zero, even though a perfect straight line can be drawn through the points. If however, the density results are taken to five places, and plotted as shown in Fig. 5.2(b), it becomes obvious that there is a relationship, and even though the results are more scattered, the correlation coefficient of 0.961, obtained by substituting the unrounded densities into equation (5.5), is considerably greater than that obtained previously.

The correlation coefficient is also influenced by scatter; the greater the scatter about a regression line, the smaller will be the value of r. If the points in Fig. 5.2 had coincided exactly with the line, the correlation coefficient would have been higher than 0.961. When the points are numerous and highly scattered, it is difficult to establish by eye whether or not the results are correlated. Thus Fig. 5.2(c) suggests there is a positive relationship, but it is not clear if it is rectilinear. Similarly, the appearance of Fig. 5.2(d) suggests that there is no correlation between x and y, but the probability of this assumption is indefinite, unless the correlation coefficient is calculated.

The basis of the correlation coefficient concept can be seen if two straight lines at right angles, and parallel to the axes, are drawn through the point representing the mean values of x and y, as shown in Fig. 5.2(c) and (d). If x is positively related to x and y, the majority of the points will be located in areas B and C, if they are negatively related, the points will lie in A and D, and if they are unrelated, the points will be uniformly located in all four areas. This is shown in Fig. 5.2(c) and (d).

For all points in areas B and C, the terms $(x - \overline{x})$ and $(y - \overline{y})$ will have the same sign, while in areas A and D they will have different signs. Therefore for positive relationships between x and y, $S(x - \overline{x})(y - \overline{y})$ will be a high positive number, and for negative relationships it will be a high negative number. For scattered results, it will be a low number of either sign. By convention, the sign of the correlation coefficient is not usually quoted.

It may seem strange that a good correlation has been established when the relative change in density (0.05%) is insignificant in comparison with a 300% increase in temperature. The reason is that the correlation process involves standardizing the raw data. This is achieved by subtracting the means from each result and dividing by the standard deviation (page 83). This brings the two sets of variables into a similar size range, each with a total of zero, and a standard deviation of one.

Equation (5.5) can be expressed in the form of equation (5.6), which is a more convenient way than equation (5.5) of calculating correlation coefficients on an electronic calculator. The symbols have been defined, and the numerator and the first term of the denominator of equation (5.6) calculated above (page 58). The second term in the denominator can be calculated in a similar way, to give equation (5.7).

$$r = \sqrt{\frac{S(xy) - S(x)S(y)/N}{(S(x^2) - S^2(x)/N)(S(y^2) - S^2(y)/N))}} \tag{5.6}$$

$$r = \sqrt{\frac{1097.47 - 989.97}{((3408.95 - 3030.72)(354.45 - 323.37))}} \tag{5.7}$$

s is the standard error of the estimate, although in much of the published QSAR work it is called the standard deviation. Standard deviation is a measure of the difference within a set of replicates, thus for a collection of values of y,

$$\text{Standard deviation} = S_y = \sqrt{\frac{S(y-\bar{y})^2}{N-1}} \tag{5.8}$$

while the standard error of the estimate is the equivalent term for the differences between the raw results (y) and those predicted from the regression equation (y_{pred}), i.e.

$$s_{xy} = \sqrt{\frac{S(y-y_{pred})^2}{N-2}} \tag{5.9}$$

The denominator has been transformed from $\surd(N-1)$ to $\surd(N-2)$ because a second degree of freedom has been lost. S_{xy} is therefore a measure of the precision with which the regression equation fits the experimental values of the dependent variable. Standard error of the estimate can be calculated more conveniently from equation (5.10),

$$s_{xy} = \sqrt{\frac{S(y^2) - S^2(xy)/S(x^2)}{N-2}} \tag{5.10}$$

so that for the data in Table 5.2,

$$s_{xy} = \sqrt{\frac{354.4469 - 1097.46^2/3408.95}{3}}$$

$$= 0.614$$

The standard error of the coefficient is the number in brackets (0.013) following the regression coefficient of equation (5.4), and indicates that if the experiment were repeated, the value of the coefficient should lie between 0.284 ± 0.013. The greater the standard error of the coefficient, the less reliable is the coefficient, and the less is the likelihood that the regression equation represents the raw data.

The confidence in the regression coefficient can be assessed by dividing the coefficient by the standard error, which in the present case is equal to 21.2. This is a high number, which suggests that the relationship is good, an opinion which can be confirmed by comparing the ratio with the Student's t value. A table of these is provided on page 179. The degrees of freedom are equal to the number of pairs of results, minus the number of variables on the right-hand side of the regression equation. In this case there is only one variable (x) on the right-hand side, so that the number of degrees of freedom (ϕ) is equal to $5-1=4$. The critical t value corresponding to a probability level of 0.01 is given as 4.60 in the table, so that since this is less than 21.2, the indications are that the chance that the coefficient does not represent a true relationship is less than 1 in 100. In other words, the relationship is highly probable. The ratio of the coefficient and the standard error of the estimate, in this case 21.2, is frequently placed below the coefficient, as shown in equation (5.4). The standard error of the intercept (0.562) is also shown in brackets. It is obtained in a similar manner to the standard error of the coefficient, and can be used in the same way. This parameter is frequently rather large, because small variations in the coefficient can bring about considerable movement in the point where the regression line crosses the x-axis, particularly when the extrapolation to zero is extensive.

The F value of the regression equation is the expression in the bottom left-hand corner of equation (5.4), and is an indication of the probability that the equation is a true relationship between the variables, and is not coincidental. The first mumber in the subscript following the symbol F, is equal to the number of independent variables on the right-hand side of the equation, in this case, 1. The second subscript is the number of sets of results ($N=5$), minus the first subscript, minus 1 degree of freedom, i.e. $5-1-1=3$. Critical F values can be obtained from tables, of the form shown on page 180. The numbers running along the top of the table represent the first number in the subscript, and those running down the left-hand side represent the second number in the subscript. Thus for $\phi_1=1$, $\phi_2=3$ and a probability of 0.01, the critical value of F is 34.1, and quoted in the format as $\alpha, 0.01=34.1$. The experimental value is given as 174.2, which is greater than 34.1, indicating that the probability that the relationship is a coincidence is less than 1 in 100. F values are discussed in more detail under analysis of variance. Calculation of F value and standard errors of coefficient and intercept are incorporated into the BASIC program for linear regression, shown in the appendices.

MULTIPLE LINEAR REGRESSION

Linear regression, as discussed so far, concerns relationships between a dependent variable (y) and an independent variable (x). Because there are only two variables the relationship involves only two dimensions, so that results can be plotted on graph paper. Sometimes more than two variables are involved, for example a dependent variable (y) can be related to two independent variables (x and z), as in equation (5.11). b_0, b_1 and b_2 are constants. This involves three dimensions, so the complete picture cannot be represented by a graph.

$$y = b_0 + b_1 x + b_2 z \tag{5.11}$$

For visual representation a three-dimensional model or diagram is required. Alternatively, one of the independent variables, for example z, can be kept constant, and the other (x) plotted against y on graph paper or the visual display of a computer. The slope (b_1) of this linear regression line through the points is described as the partial regression coefficient of y on x, and its correlation coefficient is the partial correlation coefficient of y on x. A similar correlation of y against z can be carried out, and the slope or partial regression coefficient would then be equal to b_2 in equation (5.11).

The coefficients in equation (5.11) can be calculated from the simultaneous equations (5.12) to (5.14). The symbols are as

$$S(y) = b_0N + b_1S(x) + b_2S(z) \tag{5.12}$$

$$S(xy) = b_0S(x) + b_1S(x^2) + b_2S(xz) \tag{5.13}$$

$$S(yz) = b_0S(z) + b_1S(xz) + b_2S(z^2) \tag{5.14}$$

defined for linear regression. Solution of these equations will be illustrated by reference to the results from an investigation of the carminative activities of a range of volatile compounds (Evans *et al*. 1978). All the compounds, shown in Table 5.3, involved a substituent group containing an oxygen atom linked to hydrogen, an alkyl group or an alkoxy group. A hypothesis that carminative activity is dependent on the ocatanol–water partition coefficients of the compounds, coupled with the bulkiness of the smaller group attached to oxygen, was tested by fitting the results of equation (5.15) in which b_0, b_1 and b_2 are constants, ID_{50} is the

$$\log 1/\mathrm{ID}_{50} = b_0 + b_1 \log P + b_2V_\mathrm{w} \tag{5.15}$$

concentration required to reduce the response to carbachol by 50%, P the partition coefficient and V_w, the van der Waals volume of the substituent group. The results for 26 compounds are shown in Table 5.3, together with the sums of the parameters required to solve equations (5.12) to (5.14). Substitution in equations (5.12) to (5.14) gives equations (5.16) to (5.18). These equations can be solved using

$$26b_0 + 44.79b_1 + 41.19b_2 - 33.96 = 0 \tag{5.16}$$

$$44.79b_0 + 95.55b_1 + 68.70b_2 - 67.14 = 0 \tag{5.17}$$

$$41.19b_0 + 68.70b_1 + 166.4b_2 - 41.76 = 0 \tag{5.18}$$

determinants. The procedure is highly protracted, and is more easily carried out by computer. A BASIC program for three variable regression is given in the appendices. Larger sets of equations, containing more than three variables, can be handled by MINITAB. The necessary commands are given in the appendices.

When the independent variables are related, an interaction term must be introduced into equation (5.15), to give equation (5.19).

$$\log 1/ID_{50} = b_0 + b_1 \log P + b_2V_\mathrm{w} + b_3V_\mathrm{w} \log P \tag{5.19}$$

The simplest way of solving an equation of this type is to calculate the interaction terms ($V_\mathrm{w} \log P$) beforehand, and introduce them as another independent variable.

Table 5.3 — Carminative activities, partition coefficients and molar volumes

Compound	Hindering group	log 1/ID_{50} (y)	log P (x)	V_w (z)
1. Isobutanol	H	0.775	0.74	0.22
2. *n*-Butyl acetate	$CH_3C=$	1.36	1.74	3.64
3. 1,2-Dihydroxybenzene	H	1.02	0.95	0.22
4. 1,3-Dihydroxybenzene	H	1.05	0.79	0.22
5. 1,4-Dihydroxybenzene	H	0.91	0.55	0.22
6. *o*-Cresol	H	1.64	1.95	0.22
7. *m*-Cresol	H	1.54	1.99	0.22
8. *p*-Cresol	H	1.54	1.93	0.22
9. Dibutyl ether	$CH_3(CH_2)_3$	1.23	3.06	6.51
10. Diethyl ether	CH_3CH_2	0.59	0.80	3.43
11. 3,4-Dimethylphenol	H	1.91	2.42	0.22
12. Di-isopropyl ether	$(CH_3)_2CH$	0.71	1.63	4.97
13. Di-*n*-propyl ether	$CH_3(CH_2)_2$	1.00	3.03	4.97
14. Ethyl acetate	$CH_3C=$	0.59	0.70	3.64
15. Ethylvinyl ether	CH_3CH_2	1.21	1.04	3.01
16. Eugenol	H	2.43	2.99	0.22
17. 1-Hexanol	H	1.47	2.03	0.22
18. Menthol	H	2.13	3.31	0.22
19. 2-Methoxyphenol	H	1.26	1.90	0.22
20. 4.Methoxyphenol	H	1.32	1.34	0.22
21. 1-Pentanol	H	1.11	1.16	0.22
22. 2-Phenoxyethanol	H	0.90	1.16	0.22
23. Isopropyl acetate	$CH_3C=$	0.96	1.02	3.64
24. *n*-Propyl acetate	$CH_3C=$	0.94	1.50	3.64
25. Salicylaldehyde	H	1.70	1.76	0.22
26. Thymol	H	2.66	3.30	0.22
Totals		33.96	44.79	41.19

$S(x^2) = 95.55$; $S(xz) = 68.70$; $S(xy) = 67.14$; $S(yz) = 41.76$; $S(z^2) = 166.43$.

As with linear regression analysis, additional parameters are required to support the validity of the equation. These are the same as before, but are more complicated. The more important of these are as follows.

Correlation coefficients

There are four correlation coefficients associated with equation (5.11), three linear correlation coefficients, one for each combination of two variables, i.e. r_{xy}, r_{xz} and r_{yz}, plus the coefficient of multiple regression, $r_{y,xz}$, which applies to the complete equation. It may be calculated from equation (5.20).

$$r_{y,xz} = \sqrt{\frac{r_{xy}^2 + r_{xz}^2 - 2r_{xy}r_{xz}r_{yz}}{1 - r_{xy}^2}} \tag{5.20}$$

If r_{xz} is significant, it means that the so-called independent variables are not independent, and are co-related. In this situation, one should consider ignoring either x or z, and looking for a simpler relationship. Inter-relationships between 'independent' variables will be considered in more detail in Chapter 6.

The linear correlation coefficients for the data in Table 5.3 are $r_{xy} = 0.759$, $r_{xz} = -0.450$ and $r_{yz} = 0.063$, so that the coefficient of multiple regression is given by,

$$r_{y,xz} = \sqrt{\frac{0.759^2 + 0.450^2 - 2 \times 0.759 \times -0.450 \times 0.063}{1 - 0.063^2}} \tag{5.21}$$

$$= 0.908$$

As stated before, what constitutes a satisfactory correlation coefficient is dependent on the purpose for which it is to be used, and on the nature of the raw data. An additional feature with respect to multiple regression is that for a given number of sets of data, the more variables considered, the better will the coefficient of multiple regression appear to be. If for example, there are two variables and two pairs of results, regression analysis will inevitably give a correlation coefficient of 1.000, even if the numbers are randomly chosen, because the best fit to any pair of points is a straight line. Similarly, if we are trying to relate five systems, and data on five variables are available, the more variables that are drawn into the correlation, the better will be the coefficient of multiple regression, until when all five variables are considered, the coefficient will equal one. In QSAR, it is reckoned that five extra regression points are the minimum necessary for an additional variable.

Standard error of the estimate

Basically, the standard error of the estimate in multiple regression analysis is the same as that obtained in linear regression analysis, as defined by equation (5.9), but this time y_{pred} is the value of y predicted by an equation containing two independent variables. The standard error of the estimate ($s_{y,xz}$) of an equation taking the form of equation (5.11) can be calculated from equation (5.22). r_{xy} is the linear correlation

$$s_{y,xz} = s_y \sqrt{\frac{1 - r_{xy}^2 - r_{xz}^2 - r_{yz}^2 + 2r_{xy}r_{xz}r_{yz}}{1 - r_{yz}^2}} \tag{5.22}$$

coefficient between x and y, with z constant, and so on. r_{xy}, r_{xz} and r_{yz} are sometimes termed zero order correlation coefficients. s_y is the standard deviation of y. Substitution of the results from Table 5.3 into equation (5.22) gives,

$$s_{y.xz} = 0.522\sqrt{\frac{1-0.759^2-0.450^2-0.063^2+2\times 0.759\times -0.450\times 0.063}{1-0.063^2}}$$

$$= 0.219$$

Standard errors of the coefficients and the intercept

These are calculated in similar manner to that shown for linear regression, and are displayed in the same way. However, the process is more protracted, so that the parameters are normally obtained using a computer package, as demonstrated, using MINITAB, in the appendices to this volume.

F value

This has the same meaning as in linear regression analysis, and is displayed in the same manner. An additional degree of freedom is subtracted for each additional variable. Thus if n is the number of sets of data and m the number of variables in the regression equation, the F value will be displayed as $F_{(m-1),(n-m)}$. Calculation of F values for multilinear equations is extremely protracted. A MINITAB program for calculating these is given in the appendices.

STEPWISE REGRESSION

This is performed when there is a selection of variables, and it is required to know what combination gives the best relationship with the dependent variable. The dependent variable is first regressed with each independent variable in turn, and the independent variable which alone gives the highest value of r^2 is selected. The correlation coefficient is squared to overcome the problems of negative values. In the second step, the dependent variable is regressed against the selected independent variable plus each of the rejected variables, in turn, giving a series of three-variable equations. The combination giving the highest value of r^2 is then selected, and the process repeated with each of the remaining independent variables, plus the two selected variables. The process can be continued indefinitely, within the confines of the amount of experimental data available, and the value of each additional predictor can be judged from the magnitude in the improvement in r^2.

NON-LINEAR REGRESSION

There are an infinite number of ways in which a pair of variables may be related. The simplest situation occurs when the variables are proportional, so that a plot of one variable against the other yields a straight line. Linear regression analysis can then be applied. Often, however, variables are not proportional, but are otherwise related, and alternative means must be applied to derive a mathematical formula which fits the results. The process is called curve fitting. Three procedures which can be attempted are

(a) fitting the results to a power series;
(b) fitting the results to a theoretical model;
(c) plotting a scatter diagram and recognizing a pattern.

The power series

Often a plot of one variable against another follows a regular profile which is not a straight line. In these circumstances it may be possible to obtain a relationship by fitting the results to a power series. Thus, two variables x and y may not plot to give a straight line, as expressed by equation (5.23), but

$$y = b_0 + b_1 x \tag{5.23}$$

the deviation from linearity can be resolved by introducing a squared term, as in equation (5.24).

$$y = b_0 + b_1 x - b_2 x^2 \tag{5.24}$$

The constants in equation (5.24) can be evaluated by solving the simultaneous equations (5.25) to (5.27).

$$S(y) = b_0 N + b_1 S(x) + b_2 S(x^2) \tag{5.25}$$

$$S(xy) = b_0 S(x) + b_1 S(x^2) + b_2 S(x^3) \tag{5.26}$$

$$S(x^2 y) = b_0 S(x^2) + b_1 S(x^3) + b_2 S(x^4) \tag{5.27}$$

It will be noted that equation (5.25) is the summed form of equation (5.24), and equations (5.26) and (5.27) are equations (5.25) multiplied by x and x^2 respectively. This is an empirical observation, but it helps in remembering the equations.

Equation (5.24) is an example of a parabolic or quadratic equation. The shape of the plot of a parabolic relationship depends on the signs of the coefficients, the relative values of the coefficients, and the value of the independent variable. Using equation (5.24) as an example, when t is less than one, t^2 will be less than t, so that $b_2 t^2$ will be small in comparison with $b_1 t$ when b_2 is sufficiently small in comparison with b_1. The equation will then approximate to equation (5.23), which is a straight-line equation. However, as t increases, $b_2 t^2$ will increase in comparison with $b_1 t$, and when $b_1 t = b_2 t^2$, the plot will reach a maximum, and beyond this, will decline. The resultant plot is called a parabola, an example of which is shown in Fig. 5.3. If the

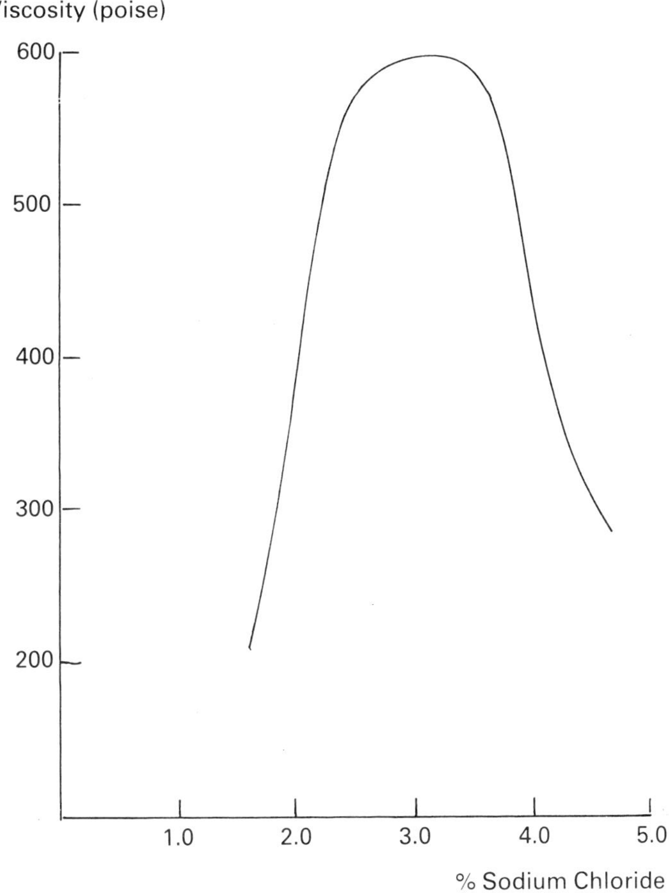

Fig. 5.3 — Viscosities of a shampoo–detergent mixture containing varying quantities of sodium chloride.

signs of its equation (5.29) are reversed, as in equation (5.28), the parabola will be inverted, and will pass through a minimum.

$$y = b_0 - b_1 x + b_2 x^2 \tag{5.28}$$

Sometimes the complete parabola fits the experimental facts, Hansch and Fujita (1964) for example, have provided theoretical and experimental evidence for a parabolic relationship between biological activity and partition coefficient, and it is well known that some detergent mixtures have viscosities which vary parabolically with the concentration of electrolyte they contain, as shown in Fig. 5.3.

The position of the maximum in a parabola is easily determined by differentiation of the equation and placing the result equal to zero. Thus, regression analysis of the data used to plot Fig. 5.3 yielded equation (5.29), in which c_{NaCl} represents the

$$\text{Viscosity (poises)} = -74.3 + 86.1\,c_{NaCl} - 14.0\,c_{NaCl}^2 \tag{5.29}$$

percentage concentration of electrolyte. Differentiation gives equation 5.31. At the maximum the slope of the graph is zero

$$\frac{\text{d.viscosity}}{\text{d.\% electrolyte}} = 86.1 - 28.0\,c_{NaCl} \tag{5.30}$$

so that the concentration of electrolyte which gives the most viscous solution is equal to 86.1/28.0 = 3.01%. The same process can be used to locate a minimum in a parabola.

Quadratic equations have been used to calculate the initial rates of enzymatic hydrolysis of some testosterone esters, *in vitro* (James *et al*. 1975). A typical plot of percentage reaction product (testosterone alcohol), shown in Fig. 5.4, is curved, and was shown by regression analysis to fit equation (5.31).

$$\%\ \text{testosterone} = 3.67 + 4.81t - 0.082t^2 \tag{5.31}$$

The line drawn through the points in Fig. 5.4 was calculated from (5.31), and can be seen to fit the data. James *et al*. (1975) used this procedure to compare hydrolysis rates. The rate of hydrolysis at the commencement of the reaction (zero rate) was estimated by differentiating, and then placing t equal to zero. Differentiation of equation (5.31) gave equation (5.32).

$$\frac{\text{d}(\%\ \text{testosterone})}{\text{d}t} = 4.81 - 0.164t \tag{5.32}$$

so that when $t = 0$,

$$\left[\frac{\text{d}(\%\ \text{testosterone})}{\text{d}t}\right]_{t=0} = 4.81\ \text{min}^{-1} \tag{5.33}$$

Curves of the type shown in Fig. 5.4 will only fit a quadratic equation when the points form part of the positive or negative curved slopes of the parabola, and in general it must be remembered that the quadratic relationship applies only to the range over which the observed results lie.

Parabolic curves are symmetrical, but sometimes skewed results are obtained.

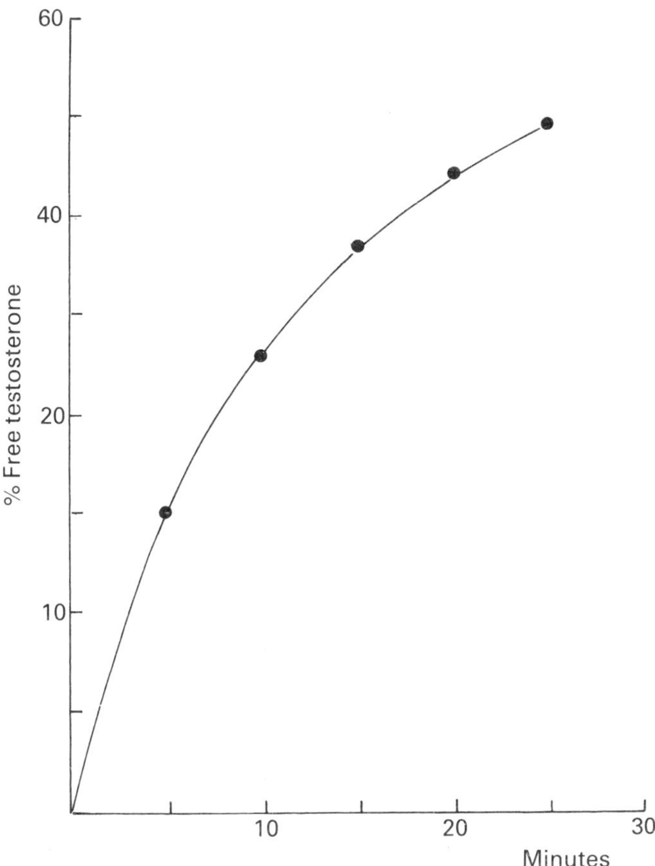

Fig. 5.4 — Enzymic hydrolysis of testosterone propionate.

The best examples of skewed parabolic curves occur when the diameters of populations of milled particles are plotted against frequency of occurrence. While naturally occurring particles give normal distribution curves, milled powders give skewed distributions. The usual treatment on such occasions is to plot log particle size against frequency, which gives a normal distribution.

James and Ng (1972) applied this principle to the time-response plots obtained in castrated rats after intramuscular injections of testosterone esters. These plots had a serrated appearance, but could be approximated to a skewed distribution curve, from which the area under the curve could be used as an approximate measure of overall androgenic response, and the position of the maximum (the time of maximum effect) used as an indication of the duration of the response. The necessary equations were obtained by correlating log biological response against time. Typical plots are shown in Fig. 5.5.

When a plot of x against y which deviates from linearity fails to fit a quadratic equation, it may be possible that a cubed term, as shown in equation (5.34), will bring about the necessary correction.

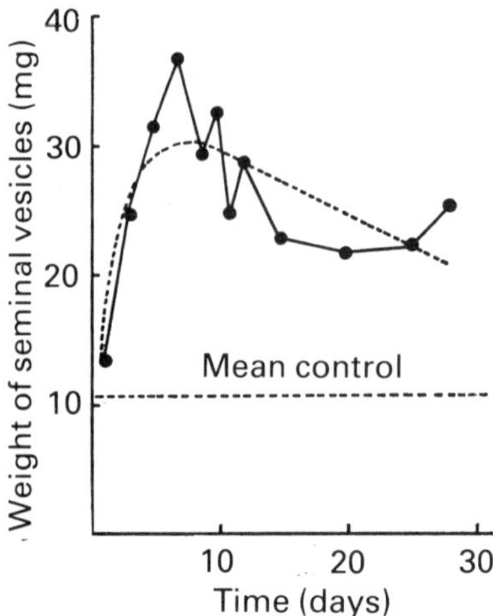

Fig. 5.5 — Time-response plots for 400 μg androstanolone valerate on seminal vesicles. —●—, experimental; – – –, calculated. (From *J. Pharm. Pharmacol.* (1972), 25 (Suppl.) 52–56.)

$$y = b_0 + b_1x + b_2x^2 + b_3x^3 \tag{5.34}$$

This is an example of a ternary equation. The classical plot for a ternary equation will pass through a maximum and a minimum, but plots deviating from this form can be obtained when roots are imaginary, or when two or more of the three roots are equal. The process can be expanded to a quartic equation, involving a term in x^4, and thence to higher powers. Eventually, when the number of pairs of results equals the number of terms on the right-hand side of the correlation equation, a perfect relationship will result. A series of terms in progressively increasing or decreasing powers of the independent variable, is called a power series.

The virial equation for unexpanded gases is probably the best known example of the power series approach. Expanded gases obey the perfect gas law equation (5.35),

$$\frac{PV}{RT} = 1 \tag{5.35}$$

where P, V and T are pressure, volume and temperature respectively, and R is the gas constant. As the volume decreases, gas molecules come closer together, and intermolecular forces become increasingly more important. As a result, the gas deviates more and more, with increasing pressure, from the relationship expressed by equation (5.35). Deviations are straightened out in the virial equation by transforming the relationship to a power series in V, as follows,

$$\frac{PV}{RT} = 1 + BV^{-1} + CV^{-2} + DV^{-3} + \dots \qquad (5.36)$$

B, C and D are virial coefficients, and are constant for a given system. As V increases, the higher power terms become progressively smaller, until the series reduces to equation (5.35), so that the number of terms necessary to completely describe the system is self limiting.

The simultaneous equations required to derive a power series equation are derived by multiplying the basic equation,

$$y = b_0 + b_1x + b_2x^2 + b_3x^3 \dots bn_nx^n \qquad (5.37)$$

progressively by $1, x, x^2 \dots x^n$, and summing, to give the equations grouped under equation (5.38).

$$\begin{aligned} S(y) &= b_0N + b_1S(X) + b_2S(X^2) \dots b_nS(X^n) \\ S(xy) &= b_0S(x) + b_1S(x^2) + \dots b_nS(x^{n+1}) \\ S(x^{n-1}y) &= b_0S(x^{n-1}) + b_1S(x^n) + \dots b_nS(x^{2n}) \end{aligned} \qquad (5.38)$$

Such exercises are too protracted to carry out manually, and it is advised that a computer be used. A program for parabolic equations is shown in the appendices, and is suitable for microcomputers. Similar programs for higher order equations can be prepared. The necessary MINITAB commands for polynomial equations are also given in the appendices.

Tests for the goodness of fit of polynomial equations present the problem that linearity cannot be established by plotting a function of one variable against a function of the other, because the terms on the right-hand side of the equation cannot be resolved into one function of the independent variable. The easiest way of testing such a relationship is to compare observed results with calculated results, either in the form of a table, or of a plot of observed results against calculated results. For a good fit, a straight line, passing through or extrapolating to the origin, should be obtained.

FITTING CURVES TO MODELS

It is sometimes possible to speculate on a mechanism upon which to base a curve-fitting exercise. This may be a well-established principle; for example, if a property of a system varies with time, the rate of change may follow first-order kinetics, so that a plot of the logarithm of the property against time is rectilinear. If this fails to give a straight line, a plot of the reciprocal of the property against time, as in second-order kinetics, might be successful. Again, a change in a property with temperature may respond to a plot of the logarithm of the property against the reciprocal of temperature, as in an Arrhenius plot. It is expedient to think deeply about the classical laws of physics and chemistry, before resorting to trial and error in fitting a curve to your data. Time can sometimes be saved by looking through the literature for systems which are similar, and for which mathematical relationships have been developed.

Careful appraisal of the system under examination, and speculation on what is happening to it, can help. As an example, release of drug from an oily base through an aqueous layer was considered to involve migration from the oily layer, accompanied by back migration from the aqueous phase as it approached saturation (Armstrong *et al.* 1988). This is reminiscent of the kinetics of opposing reactions, and consultation of standard kinetics text books revealed that these follow equation (5.39).

$$\ln\frac{Q_0 - Q_e}{Q_t - Q_e} = b_1 t \tag{5.39}$$

Q_0, Q_e and Q_t represent concentrations at zero time, equilibrium and time t respectively. This approach was tested by plotting the left-hand side of equation (5.39) against t, and found to give a straight line (Armstrong *et al.* 1988).

A mathematical analysis of the system under scrutiny can sometimes lead to a suitable graphical relationship. Higuchi (1960) examined the percutaneous absorption process from suspensions of drugs in creams and ointments, and derived equation (5.40). C_s is the concentration in the *stratum corneum*

$$C_s = K\sqrt{2ADtS_v} \tag{5.40}$$

and S_v is the solubility in the vehicle, A is the concentration of drug suspended in the base, D is the diffusion coefficient, K is the partition coefficient between *S. corneum* and base and t is time. A is usually in considerable excess of S_v and can be considered constant for a given system, as also can K and D. Resulting from this, numerous other authors have plotted drug concentration in absorption studies against the square root of time, to get a rectilinear relationship. Equations can be adapted to suit similar situations. Katz and Shaikh (1960) took Higuchi's relationship a stage further in comparing absorption rates of a collection of corticosteroids, by plotting $K\sqrt{S_v}$ against anti-inflammatory activity, and obtained a straight line.

Relationships between two variables can thus be detected by basing a process on a model. A rectilinear plot can be obtained by rearranging the model equation so that all the constants are on the right-hand side, and the variables are on the left. If the left-hand side of the equation takes the form of the ratio of one variable over the other, the graph of one variable against the other should give a straight line. If the left-hand side is the product of the two variables, then one will have to be plotted against the reciprocal of the other.

In more complicated situations, a relationship can be tested by plotting observed results against those predicted by the model equation. If the model fits the system, the plot will take the form of a straight line, passing through the origin, or extrapolating to the origin. Closeness of fit can then be tested using the procedures described under linear regression analysis.

CURVE FITTING WITHOUT MODELS

There are an infinite number of ways in which a pair of variables may be related. When they are related by simple proportion, a plot of one variable against the other will give a straight line, the equation of which can be determined by linear regression analysis, as described at the beginning of this chapter. Sometimes a relating equation can be obtained by speculating on a model mechanism, and examples of this approach have been given above. Frequently, however, the relationship is not rectilinear, and no suitable model can be found. Under such circumstances, a scatter diagram can be constructed, by plotting one variable against the other, and speculating on the type of relationship which is operating from the pattern which the points form. The operation can be carried out with pencil and graph paper, but it is easier and quicker to display the diagrams on a suitably programmed computer. Some possible equations are given in Table 5.4.

Table 5.4 — Curve types

Figure	Curve type	Equation
1a	Exponential or logarithmic	$y = b_0 b_1^x$ or $\log y = \log b_0 + x \log b_1$ or $\log y = b_0' + b_1' x$
	Where $b_0' = \log b_0$ and $b_1' = \log b_1$	
1b	Parabolic or quadratic	$y = b_0 + b_1 x + b_2 x^2$
1c	Cubic	$y = b_0 + b_1 x + b_2 x^2 + b_3 x^3$
1d	Polynomial	$y = b_0 + b_1 x + b_2 x^2 + \ldots b_n x^n$
Types 1b to 1d have been considered above.		
1e	Hyperbola	$1/y = b_0 + b_1 x$ or $y = 1/(b_0 + b_1 x)$
1f	Rectangular hyperbola	$y = 1/b_1 x$
1g	Geometric	$y = b_0^{x b_1}$ or $\log y = \log b_0 + b_1 \log x$ or $\log y = b_0' + b_1 \log x$
	Where $b_0' = \log b_0$	

It has already been explained that in first order reactions, a plot of concentration of reactant remaining against time will give an exponential plot, as shown in Fig. 5.6(a). Numerous other processes follow such relationships, so that if a process

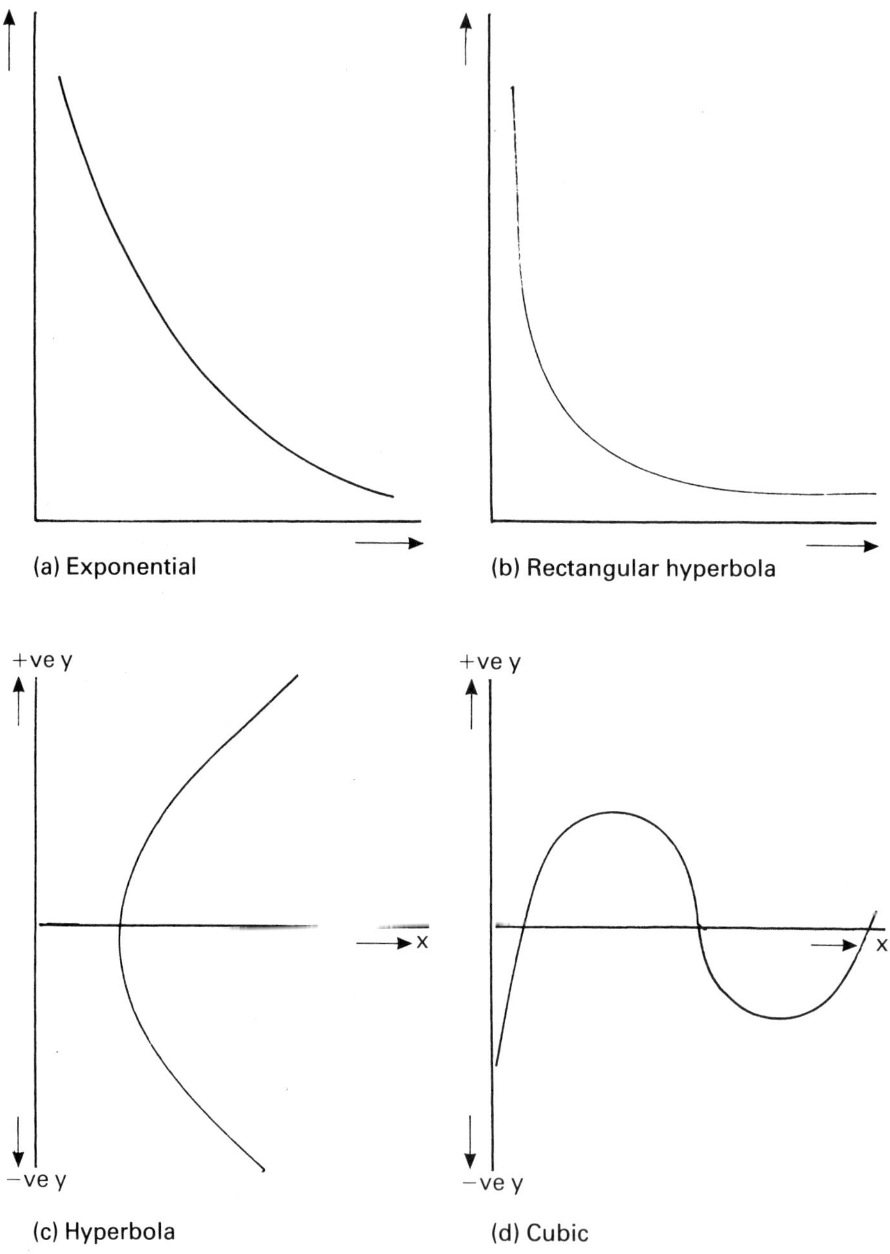

Fig. 5.6 — Curve types.

for which there is no model gives a scatter diagram which resembles Fig. 5.6(a), it is worthwhile to try plotting the variables which form the abscissa against the logarithm of the variables forming the ordinate, and if a straight line is obtained, an exponential relationship can be assumed. Sometimes however, a rectilinear logarithmic plot is not obtained, even though the scatter plot appears to be exponential. In such circumstances, the plot may follow a rectangular hyperbola, which has a similar shape to an exponential curve (Fig. 5.6(b)). Barnett and James (1979) measured the particle sizes of a wet ball-milled charge after various milling times, and obtained a plot of particle size against milling time which appeared to be exponential, but the plot of log particle size against time did not give the expected straight line. However, a plot of the reciprocal of particle size against time was rectilinear. The best fitting equation gave answers which agreed well with the observed results and a milling mechanisn based on the hyperbolic relationship was evolved. The equation for a rectangular hyperbola is given as 1f in Table 5.4. If data follow this pattern, a plot of one variable against the reciprocal of the other will give a straight line. Geometric relationships can be examined by plotting the variables on log-log paper, or by first converting the variables to their logarithms and plotting either on linear graph paper, or on the visual display of a computer.

FREE–WILSON ANALYSIS

Most investigations in quantitative structure-activity relationships (QSAR) are concerned with correlations between biological activities and physical properties. These physical properties are, in turn, dependent on the structures of the molecules involved. It is therefore logical to relate biological activity directly with chemical structure, thereby eliminating the middle ground of physical properties. Free and Wilson (1964) devised a scheme along these lines by allocating contributions to biological activities for substituent groups within the molecules.

Free and Wilson (1964) considered four analgesics with basic structure (**I**), R_1 is H or CH_3, and R_2, $N(CH_3)_2$ or $N(C_2H_5)_2$. LD_{50} values for the four compounds, in mg per 10 g, are given in Table 5.5.

Table 5.5 — LD_{50} values for four analgesic compounds

R_2	R_1			
	H	CH_3	Total	Mean
$N(CH_3)_2$	2.13	1.64	3.77	1.885
$N(C_2H_5)_2$	1.28	0.85	2.13	1.065
Total	3.41	2.49		2.950
Mean	1.705	1.245	2.950	1.475

(I)

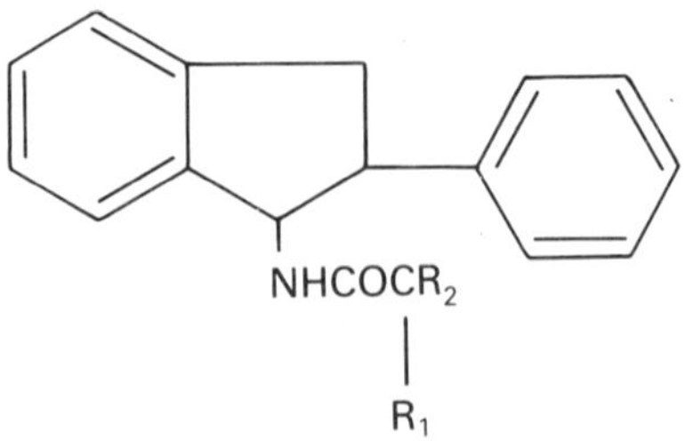

NHCOCR$_2$

R$_1$

They proposed that the biological activity of each compound was the sum of contributions of the basic chemical structure, plus contributions from each of the two substituents. Thus for the compound in which $R_1 = H$ and $R_2 = N(CH_3)_2$,

$$2.13 = \mu + a[H] + b[N(CH_3)]_2 \tag{5.41}$$

$a[\ ]$ represents the contribution of the group at position R_1, $b[\ ]$ the contribution of the group at R_2, and μ the basic contribution of the series. There are therefore three additional equations, i.e.

$$1.28 = \mu + a[H] + b[N(C_2H_5)]_2 \tag{5.42}$$

$$1.64 = \mu + a[CH_3] + b[N(CH_3)]_2 \tag{5.43}$$

$$0.85 = \mu + a[CH_3] + b[N(C_2H_5)]_2 \tag{5.44}$$

Equations (5.41) to (5.44) represent four simultaneous equations with five unknowns. The overall average (1.475) was substituted for μ, and the mean value for each substituent minus μ for the other terms in equations (5.41) to (5.44). Thus $a[H] = 1.705 - 1.475 = +0.23$, whilst $a[CH_3] = 1.245 - 1.475 = -0.23$ and $b[N(CH_3)]_2$ and $b[N(C_2H_5)_2]$ are $+0.41$ and -0.41 respectively. The virtue of this approach is that there is symmetry between the terms in each pair, which means that there are only three unknowns, but there are four ID_{50}s available.

Free and Wilson (1964) used the inhibitory activities against *S. aureus* of ten tetracycline derivatives with eight substituent groups and three positions of attachment, to demonstrate their technique. This is described fully in their paper, and elsewhere (James 1988). As a variation, the inhibiting action of chlorophenols on the biodegradation of phenol (Beltrame *et al.* 1984) will be used here as an example. Concentrations in mmol/l, producing 50% inhibition, are shown in Table 5.6. The overall mean IC_{50} is equal to the total IC_{50} divided by the number of compounds, i.e. 5.271/19 = 0.2774. There are fourteen compounds substituted by chlorine in the 2-position, giving a mean IC_{50} of 0.248, and five substituted by hydrogen in the 2-

Table 5.6 — Inhibiting action of chlorophenols on biodegradation of phenol

Compound	Inhibiting action (IC_{50}) (mmol/l)	
	Observed	Predicted
2-Chloro	0.812	0.526
3-Chloro	0.525	0.468
4-Chloro	0.552	0.480
2,3-Dichloro	0.338	0.356
2,4-Dichloro	0.292	0.368
2,5-Dichloro	0.308	0.342
2,6-Dichloro	0.400	0.428
3,4-Dichloro	0.262	0.310
3,5-Dichloro	0.358	0.285
2,3,4-Trichloro	0.139	0.198
2,3,5-Trichloro	0.113	0.172
2,3,6-Trichloro	0.199	0.258
2,4,5-Trichloro	0.120	0.184
2,4,6-Trichloro	0.213	0.270
3,4,5-Trichloro	0.099	0.126
2,3,4,5-Tetrachloro	0.088	0.014
2,3,4,6-Tetrachloro	0.175	0.100
2,3,5,6-Tetrachloro	0.191	0.074
Pentachloro	0.087	− 0.084

position, with a mean value of 0.359. The mean of the means is (0.248 + 0.359)/2 = 0.304, so that the Free–Wilson constants are

$$2\text{-Cl} = 0.248 - 0.304 = -0.056; \qquad 2\text{-H} = 0.359 - 0.304 = +0.056$$

Constants for the other substituents are given in Table 5.7. Using this information for 2-chlorophenol,

$$[2\text{-Cl}] = 0.277 - 0.056 + 0.085 + 0.079 + 0.092 + 0.049 = 0.526$$

Predicted IC_{50}s, using this procedure, are given in Table 5.6. The predictions are not very good in this example, but they illustrate the procedure. The usefulness of the method lies in the fact that after looking at about ten selected compounds from the list, a rough idea of the efficacy of the remaining compounds can be obtained.

Table 5.7 — Free–Wilson constants for inhibiting actions of chlorophenols

Substituent group	Constant	Substituent group	Constant	Substituent group	Constant
2-Cl	−0.056	3-Cl	−0.085	4-Cl	−0.079
2–H	+0.056	3-H	+0.085	4-H	+0.079
5-Cl	−0.092	6-Cl	−0.045		
5-H	+0.092	6-H	+0.045		

REFERENCES

Armstrong, N. A., Griffiths, H. A. and James, K. C. (1988) An *in vitro* model to simulate drug release from oily media, *Int. J. Pharm.* **41** 115–119.

Barnett, M. I. and James, K. C. (1979) A quantitative evaluation of size reduction in wet ball milling, *Drug. Devel. Ind. Pharm.* **5** 63–78.

Beltrame, P., Beltrame, P. L. and Carniti, P. (1984) Inhibiting action of chloro- and nitro-phenols on biodegradation of phenol: a structure toxicity relationship, *Chemosphere* **13** 3–9.

Evans, B. K., James, K. C. and Luscombe, B. K. (1978, 1979) Quantitative structure–activity relationships and carminative activity, *J. Pharm. Sci.* **67** 277–278; **68** 370–371.

Free, S. M. and Wilson, J. W. (1964) A mathematical contribution to structure-activity studies, *J. Med. Chem.* **7** 395–399.

Gebre-Mariam, T. (1988) Drug migration in soft capsules, PhD Thesis, University of Wales.

Hansch, C. and Fujita, T. (1964) π-δ-π Analysis. A method of correlation of biological activity and chemical structure, *J. Am. Chem. Soc.* **86** 1616–1626.

Higuchi, T. (1960) Physical chemical analysis of percutaneous absorption, *J. Soc. Cos. Chem.* **11** 85–97.

James, K. C. (1988) In *Introduction to the Principles of Drug Design 2 edn* (ed. Smith, H. J.), Butterworth, London, pp. 256–258.

James, K. C. and Ng, C. T. (1972) A method of interpreting time-response curves, *J. Pharm. Pharmacol.* **25** (Suppl.) 52–56P.

James, K. C., Nicholls, P. J. and Richards, G. T. (1975) Correlation of androgenic activities of the lower testosterone esters in rat, with R_m values and hydrolysis rates, *Eur. J. Med. Chem.* **10** 55–58.

Katz, M. and Sheikh, Z. I. (1960) Percutaneous corticosteroid absorption correlated to partition coefficients, *J. Pharm. Sci.* **54** 591–594.

ADDITIONAL READING

Spiegel, M. R. (1961) *Theory and Problems of Statistics*. Schaum Publishing Co., New York.

6

Multivariate analysis

INTRODUCTION TO MULTIVARIATE ANALYSIS

Regression involves relationships between one dependent variable and one or more independent variables. Multivariate analysis looks for interdependence among all variables. It considers random variables collectively, and asseses which variables are related and which are not. The subject frequently involves the use of matrices.

A matrix is a rectangular array of numbers, traditionally bounded by square brackets. It is often convenient to display raw data in this form, for example, by removing all the information outside the brackets, Table 6.1 is transformed into a

Table 6.1 — Analytical profiles of samples of olive oil

Sample	Acid value	Iodine value	Refractive index	Saponification value	Weight per ml
A	1.0	79	1.469	192	0.911
B	1.4	82	1.470	193	0.911
C	1.2	88	1.471	192	0.913
D	1.5	83	1.468	195	0.912
E	1.3	85	1.470	193	0.913
Mean	1.280	83.40	1.4696	193.0	0.912
St. Dev.	0.192	3.36	0.0011	1.225	0.001

matrix. Each number in the array is called an element, each set of elements running along a matrix is a row, and each vertical set of elements is a column. Because Table 6.1 contains 5 rows and 5 columns, it is a (5×5) matrix, and because the number of rows equals the number of columns, the matrix is said to be square. A row enclosed in square brackets, for example, [1.0 79 1.469 192 0.911], is called a row vector, and a column enclosed in square brackets, for example,

$$\begin{bmatrix} 1.0 \\ 1.4 \\ 1.2 \\ 1.5 \\ 1.3 \end{bmatrix}$$

is a column vector.

Matrices can be subjected to mathematical manipulations, such as addition and multiplication, but the procedures involve different rules from classical algebra. The manipulations relevant to the multivariate methods described in this volume will be demonstrated later in this chapter. For additional information, the reader is referred to the references at the end of this chapter.

A matrix displaying only raw data, such as Table 6.1, rarely conveys any obvious information with respect to relationships between variables. Multivariate analysis usually involves transformation of raw data to a matrix in which relationships are more easily identified. The best known of these are the distance matrix, the covariance matrix and the correlation matrix.

DISTANCE MATRIX

Table 6.1 represents the analytical profiles of five samples of olive oil. If any two columns are taken from this matrix, their elements can be plotted against each other on a piece of graph paper, and will give a visible impression of the relationship between the two properties represented. The distance between any two points on this graph can be calculated, using the Pythagoras theorem, for example, for the elements of the first two rows of the first two columns,

$$\text{Distance of separation} = \sqrt{(82-79)^2 + (1.4-1.0)^2}$$

$$= 3.027 \tag{6.1}$$

Similarly, the elements of the first three columns can be plotted as a three-dimensional model, and the distance of separation, in space, of the points representing the first two rows would be

$$\sqrt{(1.4-1.0)^2 + (82-79)^2 + (1.470-1.469)^2} = 3.027 \tag{6.2}$$

Four columns of the matrix cannot be represented pictorially, because there are only three dimensions in visual space, but by using the method of calculation, the distance between two points in four-dimensional space can be calculated. The process can be extended to an infinite number of dimensions, with increasing numbers of columns.

The use of distances can be illustrated by assuming it is required to select the sample of olive oil whose properties are nearest to sample A. The results in Table 6.1 cannot be compared directly, because they have widely different orders of magnitude, for example, weights per ml are just below unity, while saponification values are nearly 200. Calculations would therefore be heavily weighted in favour of saponification values. This is shown by the calculations carried out above, in which the two- and three-dimensional distances given in equations (6.1) and (6.2) are identical. This is because $(1.470-1.469)^2=10^{-6}$ is negligible in comparison with $(82-79)^2=9.0$. The elements of the matrix must therefore be standardized, so that they are all of equal importance. This is done by subtracting the mean of the column from each of its elements, and dividing each result by the column standard deviation. Means and standard deviations are given in Table 6.1, so that the standardized acid value for A, for example is $(1.0-1.28)/0.192=-1.456$. Standardized results are shown in Table 6.2. Typical of standardized results, the sum of the elements in each column is zero, and the standard deviation is 1.000. The distance between A and B in 5-dimensional space can now be calculated as,

Table 6.2 — Standardized values for olive oil samples

Sample	Acid value	Iodine value	Refractive index	Saponifi- value	Weight per ml
A	−1.456	−1.309	−0.526	−0.816	−1.000
B	0.624	−0.416	0.351	0.000	−1.000
C	−0.416	1.368	1.228	−0.816	1.000
D	1.144	−0.199	−1.403	1.633	0.000
E	0.104	0.476	0.351	0.000	1.000

$$\sqrt{(-1.456-0.624)^2+(-1.309+0.416)^2+(-0.526-0.351)^2+(-0.816-0.000)^2+(-1.0+1.0)^2}=2.561 \qquad (6.3)$$

The complete data are shown in Table 6.3, which is the distance matrix for the data,

Table 6.3 — Distance matrix for olive oil samples

	A	B	C	D	E
A	—	2.561	3.915	3.993	3.325
B	2.561	—	3.114	3.178	2.251
C	3.915	3.114	—	4.309	1.581
D	3.993	3.178	4.309	—	2.860
E	3.325	2.251	1.581	2.860	—

and indicate that B, which is closest to A, has an analytical profile which is nearest to that of A. Distances calculated in this way are often called Euclidian distances. It will be noted that the numbers below the leading diagonal are a mirror image of those above. For this reason, half of the matrix is usually omitted. This procedure will sometimes be used below.

COVARIANCE MATRIX

The variance of a column of elements is the square of the standard deviation of the elements. It is equal to the sum of the squares of the differences between each element and the mean of all the elements, divided by the number of elements, minus one, as expressed in equation (6.4).

$$\text{Variance } (V) = \frac{S(x-\bar{x})^2}{N-1} \tag{6.4}$$

$\bar{x}$ is the mean value, and x represents the individual values in the column. N is the number of elements. Sometimes in multivariate analysis N is used, rather than $N-1$. However, for practical purposes it is more convenient to use $N-1$, to bring results into line with normal statistical practice. It makes no difference in the long run which denominator is used, provided it is used consistently.

Variance is more easily calculated from equation (6.5).

$$V = \frac{S(x^2) - S^2(x)/N}{N-1} \tag{6.5}$$

$S(x_2)$ is the sum of the squares of all the elements, and $S^2(x)/N$ is the square of the sum of all the elements divided by the number of elements.

Table 6.4 gives the results for a quantitative structure–activity relationship exercise, and shows the androgenic activities of five compounds, together with some of the physical properties of these compounds (James *et al.* 1975). The properties and activities are explained in the legend accompanying the table. As such, the numbers in the matrix convey very little to the reader, but conversion to a covariance matrix reveals more information. For the first column of the matrix in Table 6.4,

$$S(x^2) = 1.63^2 + 2.04^2 + 2.70^2 + 2.96^2 + 2.84^2 = 30.936$$

and

$$\frac{S^2(x)}{N} = \frac{(1.63 + 2.04 + 2.70 + 2.96 + 2.84)^2}{5} = 29.622$$

therefore

$$V = \frac{30.936 - 29.622}{5-1} = 0.329$$

The covariance between a column of elements (x) and a column of elements (y) is given by equation (6.6), but can be calculated more easily using equation (6.7).

Table 6.4 — Androgenic activities and QSAR parameters of some testosterone esters (James *et al.*, 1975)

Ester	Log overall androgenic response (log OAR)	Log catalytic constant (log k_c)	R_m	E_s
Formate	1.63	1.27	0.58	0.00
Acetate	2.04	1.48	0.46	−1.24
Propionate	2.70	2.00	0.11	−1.58
Butyrate	2.96	2.09	−0.09	−1.60
Valerate	2.84	2.06	−0.26	−1.63

Overall Androgenic response represents the area under the curve obtained when the weights of prostate plus seminal vesicles of castrated rats were plotted against time since dosing.
Catalytic constant is the rate constant for the hydrolysis of the esters, *in vitro*, with standardized liver homogenate.
R_m is a chromatographic parameter derived from R_f value, and logarithmically related to log partition coefficient.
E_s is a parameter related to the bulkiness of the ester group.

$$\text{Covariance } (c_{xy}) = \frac{S(y-\bar{y})(x-\bar{x})}{N-1} \tag{6.6}$$

$$c_{xy} = \frac{S(xy) - S(x)S(y)/N}{N-1} \tag{6.7}$$

(xy) is the sum of the products of x and y, and $S(x)S(y)$ the products of the sums of x and y. N now represents the number of pairs of elements, x and y. Thus for the first two columns of the matrix in Table 6.4,

$$S(xy) = 1.63 \times 1.27 + 2.04 \times 1.48 + 2.70 \times 2.00 + 2.96 \times 2.09 +$$

$$2.84 \times 2.06 = 22.526$$

and

$$\frac{S(x)S(y)}{N} = \frac{(5 \times 2.43)(5 \times 1.78)}{5} = 21.627$$

(2.43 and 1.78 are the mean values of x and y respectively). Covariance between the elements in column 1 and 2 is thus

$$\frac{22.526 - 21.627}{5-1} = 0.225$$

The precise answer obtained to this calculation varies with the number of significant figures taken. The complete covariance matrix is shown in Table 6.5.

Table 6.5 — Covariance matrix to Table 6.4

	log OAR	log k_c	R_m	E_s
log OAR	0.329	0.225	−0.193	−0.359
log k_c		0.143	1.869	−0.232
R_m			0.127	0.442
E_s				0.483

The principal virtue of covariance matrices is that they are always square matrices, even when the matrices from which they have been derived are not. The importance of this is that several parameters associated with multivariate analysis, for example determinants, eigenvalues and eigenvectors, can only be calculated for square matrices.

CORRELATION MATRIX

Table 6.6 is the standardized form of the data given in Table 6.4.

Table 6.6 — Standardized values from Table 6.4

	log OAR	log k_c	R_m	E_s
Formate	−1.4028	−1.3475	1.1788	1.7418
Acetate	−0.6875	−0.7926	0.8420	−0.0432
Propionate	0.4641	0.5813	−0.1403	−0.5326
Butyrate	0.9178	0.8191	−0.7017	−0.5614
Valerate	0.7084	0.7398	−1.1788	0.6046

Calling the elements in the first column x,

$$S(x) = -1.4028 - 0.6875 + 0.4641 + 0.9178 + 0.7084 = 0.000$$

and

$$S(x^2) = (-1.4028)^2 + (-0.6875)^2 + (0.4641)^2 + (0.9178)^2 + (0.7084)^2 = 4.0000$$

therefore

$$\text{Variance of } x = \frac{4.0000 - 0.0000}{5 - 1} = 1.0000$$

Thus the sum is zero and the variance is unity. This is characteristic of standardized data. Since $S(x) = S(y) = 0$, $\overline{X}$ and $\bar{y}$ will also be zero, so that substitution from columns 1 and 2 of Table 6.6 into equation (6.6) yields

$$c_{xy} = (-1.4028 \times -1.3475) + \ldots (0.7084 \times 0.7398) = 0.995$$

The covariance matrix of the standardized values given in Table 6.6 is shown in Table 6.7, and displayed in cross-reference form, in the same way as Table 6.3. The

Table 6.7 — Correlation matrix for data in Table 6.4

$$\begin{bmatrix} 1.000 & 0.995 & -0.944 & -0.901 \\ & 1.000 & -0.946 & -0.882 \\ & & 1.000 & 0.800 \\ & & & 1.000 \end{bmatrix}$$

identical result is obtained by linearly regressing each column of elements, in turn, with the other columns, and displaying the correlation coefficients. The table can therefore also be described as a correlation matrix, and this title is most commonly given.

Table 6.7 reveals that the logarithm of the catalytic constant is rectilinearly related to the logarithm of overall androgenic response ($r = 0.995$), but a relationship between log OAR and R_m ($r = 0.944$) is also indicated. However, the intersection of the R_m column with the log k_c row ($r = 0.946$) suggests that this may be explained by relationships between R_m and log k_c. Further tests are necessary to resolve these problems, and will be described as they arise in the text.

DETERMINANTS

Determinants are expressions associated with simultaneous equations, and were in use before matrices were invented. Every square matrix has its determinant, which has a specific value characteristic of that matrix. To understand the calculation and importance of determinants, it is necessary to consider the solution of simultaneous equations.

The simultaneous equations (6.8) and (6.9) would traditionally be solved by multiplying the terms in equation (6.8) by 2, and the terms in equation (6.9) by 4, giving equations (6.10) and (6.11) respectively.

$$4x + y = 8 \tag{6.8}$$

$$2x + 3y = 12 \tag{6.9}$$

$$8x + 2y = 16 \tag{6.10}$$

$$8x + 12y = 48 \tag{6.11}$$

Subtraction of equation (6.10) from equation (6.11) eliminates the first term, giving a value for y of 3.2. This in turn can be substituted into equation (6.9) to give a value of 1.2 for x. The procedure involving the terms on the left-hand side is thus:

$$(4 \times 2) + (4 \times 3) - (2 \times 4) - (2 \times 1)$$

which reduces to $(4 \times 3) - (2 \times 1) = 10$. The process can be expressed in the form

$$\begin{bmatrix} 4 & 1 \\ 2 & 3 \end{bmatrix}$$

in which the 4 and 1 are the coefficients of x and y in equation (6.8) and 2 and 3 are the corresponding coefficients in equation (6.9). The vertical lines on each side of the array symbolizes that the expression is the determinant of the matrix:

$$\begin{bmatrix} 4 & 1 \\ 2 & 3 \end{bmatrix}$$

The determinant is solved by subtracting the product of the first element in the second row and the second element in the first row from the product of the first element in the first row and the second element in the second row. This example is a second order determinant because it has 2 rows and 2 columns. It evaluates to $(4 \times 3) - (2 \times 3) = 10$, which is the determinant of that particular matrix. The use of determinants in solving simultaneous equations is shown in the appendices to this volume.

Equations (6.8) and (6.9) are equations of straight lines, and if plotted, would intersect at $x = 1.2$ and $y = 3.2$, as shown in Fig. 6.1, the solution of the two equations. However, if equation (6.9) is solved in conjunction with equation (6.12), rather than equation (6.8), then multiplication of the terms in equation (6.9) by 4 and those in equation (6.12) by 2 would give identical equations, equations (6.13).

$$4x + 6y = 24 \qquad (6.12)$$

$$8x + 12y = 48 \qquad (6.13)$$

Equations (6.9), (6.12) and (6.13) give the same line in Fig. 6.1, because in all three equations if x is set equal to zero, $y = 4$, and if y is set equal to zero, $x = 6$. Put in another way, if equation (6.12) is divided through by 2, or equation (6.13) by 4, then equation (6.9) results. These relationships occur because the coefficients in x and y in equation (6.9) are proportional to those in equation (6.12).

Expressing these observations in terms of matrices, the matrix corresponding to equations (6.9) and (6.12) is

$$\begin{bmatrix} 2 & 3 \\ 4 & 6 \end{bmatrix}$$

and will have a determinant of $(2 \times 6) - (4 \times 3) = 0$. This is an example of the generalization that a matrix in which the elements in the columns are proportional will have a determinant of zero. This provides a test for proportionality between columns in a matrix. It can be argued that, for the example given, proportionality is

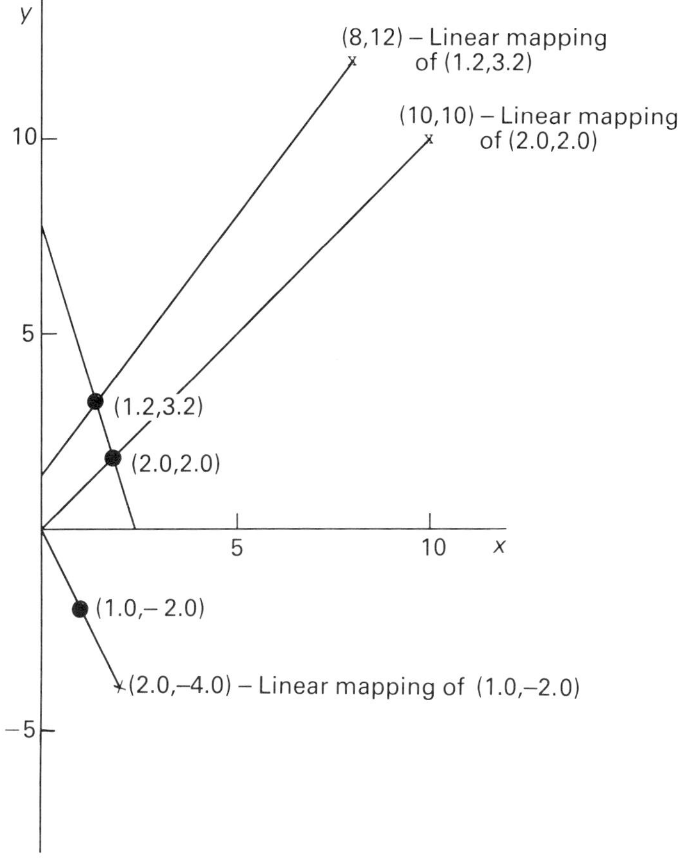

Fig. 6.1 — Linear mapping of the matrix $\begin{bmatrix} 4 & 1 \\ 2 & 3 \end{bmatrix}$.

obvious, and can be detected simply by scrutiny of the numbers in the equations. This is true for a second order matrix, which has been used here as a simple example, but as the order of the matrix increases, the value of the test becomes evident.

The value of a (3×3) determinant to establish relationships between variables can be illustrated by using the results in Table 6.8. This gives the diffusion coefficients of 4-hydroxy-benzoic acid in three gelatin gels, A, B and C, together with the microviscosities and macroviscosities of the gels (Armstrong *et al.* 1987). It was required to know which type of viscosity influenced diffusion. Simple observation of the results is all that is needed to give the answer to this question, so that statistics need not be used to establish relationships in such cases, but by looking at such a simple situation, we can see how the methodology can be applied to more complicated problems.

A matrix of the standardized values taken from Table 6.8 is shown in Table 6.9. For the diffusion coefficient of gel A,

Table 6.8 — Influence of viscosity on the migration of 4-hydroxybenzoic acid through glycerogelatin gels

Sample	Diffusion coefficient ($mm^2 h^{-1}$)	Microscopic viscosity ($Nm^{-2} \times 10^3$)	Macroscopic viscosity ($Nm^{-2} \times 10^3$)
A	0.021	13.30	2.20
B	0.040	6.52	20.2
C	0.027	10.86	26.8
Mean	0.0293	10.227	16.400
Standard deviation	0.0097	3.434	12.733

Table 6.9 — Standardized matrix obtained from data in Table 6.8

$$\begin{bmatrix} -0.8557 & 0.8949 & -1.1152 \\ 1.0982 & -1.0794 & 0.2984 \\ -0.2402 & 0.1844 & 0.8168 \end{bmatrix}$$

$$\text{Standardized result} = \frac{0.021 - 0.0293}{0.0097} = -0.8557$$

and so on.

The calculation of the determinant of this matrix is shown in equation (6.14). Each element in the first row is multiplied, in turn, by the determinant of the (2 × 2) matrix whose elements are neither in the same row nor the same column as the first row element. The (2 × 2) matrix is called the minor of its first row element. The result obtained with the second element in the first row is subtracted from the sum of the corresponding results for the first and third elements in the first row.

$$\begin{vmatrix} -0.8557 & 0.8949 & -1.1152 \\ 1.0982 & -1.0794 & 0.2984 \\ -0.2402 & 0.1844 & 0.8168 \end{vmatrix} = -0.8557 \begin{vmatrix} -1.0794 & 0.2984 \\ 0.1844 & 0.8168 \end{vmatrix}$$

$$-0.8949 \begin{vmatrix} 1.0982 & 0.2984 \\ -0.2402 & 0.8168 \end{vmatrix} + -1.1152 \begin{vmatrix} 1.0982 & -1.0794 \\ -0.2402 & -0.1844 \end{vmatrix}$$

$$= -0.8557 \times -0.9367 - 0.8949 \times 0.9687 + -1.1152 \times -0.0568$$

$$= 0.8015 - 0.8669 + 0.0633 = -0.0021 \tag{6.14}$$

The directions of the signs between the second order determinants follow logically from the method of solving simultaneous equations. The order in which the elements are taken is also important. The columns must be represented in the lower order determinants in the same order as they appear in the original matrix.

The determinant of the above (3×3) matrix, ignoring the sign, is a very small number, signifying that at least two of the columns are related to each other. The number and nature of the columns which are related can be assessed by calculating the determinants of the second order matrices. Of the three second order determinants involving diffusion and microviscosity, the largest is 0.121, ignoring the sign, while of the six involving macroviscosity, the smallest is 0.877, indicating that macroviscosities are not related to either of the other two variables. The theoretical value of zero is not obtained with any of the determinants, because of experimental scatter.

Determinants are tedious to calculate, and if required routinely are more conveniently obtained by computer. A program in BASIC, suitable for use with a microcomputer, can be found in the appendices to this volume.

The determinant of a (4×4) matrix is obtained by multiplying each element in the first row by the determinant of the (3×3) matrix with which it shares with neither a row nor a column, as shown in Table 6.10. A (4×4) matrix can be split up into progressively smaller units, each with its own determinant. The covariance matrix of Table 6.7, for example, can be split up into four (3×3) matrices and six (2×2) matrices. These are shown in Table 6.10, together with their determinants. The determinant of the (4×4) matrix is small (0.0038), and roughly equal to those of the (3×3) matrices involving the first and second columns of Table 6.7 (0.0023 and 0.0010). In contrast, the (3×3) matrices not involving both of the first two columns (log OAR and log k_c) have determinants which are about 10 times greater (0.023 and 0.018). Similarly the (2×2) matrix of Table 6.10, involving the first and second columns of Table 6.7, has a determinant of 0.010, compared with 0.105, to 0.360 for the other (2×2) matrices. The relative values of the determinants in Table 6.10 therefore confirm the conclusion gained from the correlation matrix (Table 6.7), that the rate of hydrolysis is the critical factor controlling the biological activities of the esters.

EIGENVALUES AND EIGENVECTORS

Equations (6.8) and (6.9) can be regarded in another way, namely that for the coefficients, 4, 1, 2 and 3, and for the solution $x = 1.2$ and $y = 3.2$, there are only two possible constants, namely 8 and 12. If we call these constants x' and y' respectively, they can be evaluated as follows,

$$x' = (4 \times 1.2) + (1 \times 3.2) = 8 \tag{6.15}$$

$$y' = (2 \times 1.2) + (3 \times 3.2) = 12 \tag{6.16}$$

These equations are represented in matrix algebra by equation (6.17),

$$\begin{bmatrix} x' \\ y' \end{bmatrix} = \begin{bmatrix} 4 & 1 \\ 2 & 3 \end{bmatrix} \begin{bmatrix} 1.2 \\ 3.2 \end{bmatrix} \tag{6.17}$$

Table 6.10 — Fourth order correlation matrix and its determinants, calculated from the data in Table 6.7

Second order matrices

$$\begin{array}{c|cc|c} & \log \mathrm{OAR} & \log k_c & \\ \log \mathrm{OAR} & 1.000 & 0.995 & \\ \log k_c & 0.995 & 1.000 & = 0.010 \end{array} \qquad \begin{array}{c|cc|c} & \log k_c & R_m & \\ \log k_c & 1.000 & 0.946 & \\ R_m & 0.946 & 1.000 & = 0.105 \end{array}$$

$$\begin{array}{c|cc|c} & R_m & E_s & \\ R_m & 1.000 & 0.800 & \\ E_s & 0.800 & 1.000 & = 0.360 \end{array} \qquad \begin{array}{c|cc|c} & \log \mathrm{OAR} & R_m & \\ \log \mathrm{OAR} & 1.000 & 0.944 & \\ R_m & 0.944 & 1.000 & = 0.108 \end{array}$$

$$\begin{array}{c|cc|c} & \log k_c & E_s & \\ \log k_c & 1.000 & 0.882 & \\ E_s & 0.882 & 1.000 & = 0.222 \end{array} \qquad \begin{array}{c|cc|c} & \log \mathrm{OAR} & E_s & \\ \log \mathrm{OAR} & 1.000 & 0.901 & \\ E_s & 0.901 & 1.000 & = 0.118 \end{array}$$

Third order matrices

$$\begin{array}{c|ccc|l} & \log \mathrm{OAR} & \log k_c & R_m & \\ \log \mathrm{OAR} & 1.000 & 0.995 & 0.944 & 1.000 \times 0.105 - \\ \log k_c & 0.995 & 1.000 & -0.946 & = -0.995 \times 0.108 + \\ R_m & -0.944 & -0.946 & 1.000 & +0.994 \times 0.0027 \end{array}$$
$= 0.0010$

$$\begin{array}{c|ccc|l} & \log \mathrm{OAR} & \log k_c & E_s & \\ \log \mathrm{OAR} & 1.000 & 0.995 & -0.901 & 1.000 \times 0.222 - \\ \log k_c & 0.995 & 1.000 & -0.882 & = \ 0.995 \times 0.200 + \\ E_s & -0.901 & -0.882 & 1.000 & -0.901 \times 0.023 \end{array}$$
$= 0.0023$

$$\begin{array}{c|ccc|l} & \log \mathrm{OAR} & R_m & E_s & \\ \log \mathrm{OAR} & 1.000 & 0.944 & 0.901 & 1.000 \times 0.36 - \\ R_m & 0.944 & 1.000 & 0.800 & = -0.944 \times -0.223 + \\ E_s & 0.901 & 0.800 & 1.000 & -0.901 \times 0.146 \end{array}$$
$= 0.018$

$$\begin{array}{c|ccc|l} & \log k_c & R_m & E_s & \\ \log k_c & 1.000 & -0.946 & -0.882 & 1.000 \times 0.36 - \\ R_m & -0.946 & 1.000 & 0.800 & = -0.946 \times -0.240 + \\ E_s & -0.882 & 0.800 & 1.000 & 0.882 \times 0.125 \end{array}$$
$= 0.023$

Fourth order matrix

$$\begin{array}{c|cccc|} & \log \mathrm{OAR} & \log k_c & R_m & E_s \\ \log \mathrm{OAR} & 1.000 & 0.995 & -0.944 & -0.901 \\ \log k_c & 0.995 & 1.000 & -0.946 & -0.882 \\ R_m & -0.944 & -0.946 & 1.000 & 0.800 \\ E_s & -0.901 & -0.882 & 0.800 & 1.000 \end{array}$$

$= 1.000 \times 0.023 - 0.995 \times 0.018 + -0.944 \times 0.0023 -$
$- 0.901 \times 0.0010 = 0.0038$

which means that x' and y' are obtained by multiplying the column vector

$$\begin{bmatrix} 1.2 \\ 3.2 \end{bmatrix} \text{ by the } (2 \times 2) \text{ matrix } \begin{bmatrix} 4 & 1 \\ 2 & 3 \end{bmatrix}$$

Comparison of equation (6.17) with equations (6.15) and (6.16) reveals the matrix algebra procedure for multiplying a column vector containing two elements by a (2×2) matrix. Thus, the x coordinate is equal to the product of the first element in the first row of the (2×2) matrix and the top element of the column vector, plus the product of the second element of the top row of the (2×2) matrix and the bottom element of the column vector. The y coordinate is the product of the first element of the bottom row of the (2×2) matrix and the top element of the column vector, plus the product of the second element of the bottom row of the (2×2) matrix and the bottom column vector. The procedure is shown diagramatically in Fig. 6.2, and yields $x' = 8$ and $y' = 12$.

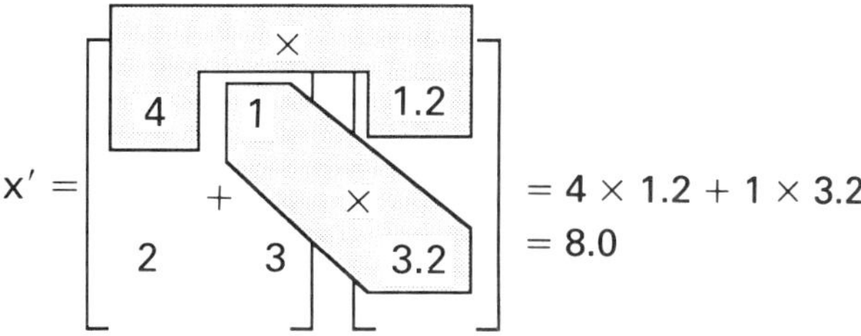

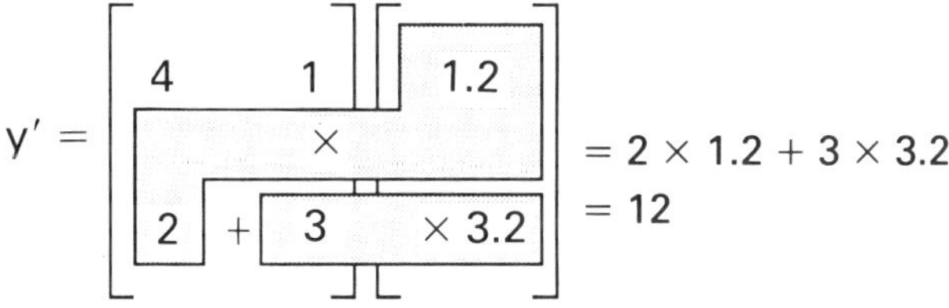

Fig. 6.2 — Multiplication of matrices.

What has been described above can be expressed in terms of a plot of x against y on graph paper, as shown in Fig. 6.1. Any point (x, y) can be related to a second point (x', y') through the general equations (6.18) and (6.19), just as the point (1.2, 3.2) is related to the point (8, 12) through equations (6.1) and (6.2). For specific values of a, b, c, d, x and y, there is only one possible pair of values of x' and y'. Thus is $x = 1.2$

and $y = 3.2$, there are only two possible values for the left-hand side of equations (6.18) and (6.19), namely

$$x' = ax + by \tag{6.18}$$

$$y' = cx + dy \tag{6.19}$$

8 and 12, or alternatively, $x' = 8$ and $y' = 12$ is the only possible solution to equation (6.17). These values are said to be a linear mapping of $x = 1.2$ and $y = 3.2$, and the matrix

$$\begin{bmatrix} 4 & 1 \\ 2 & 3 \end{bmatrix}$$

Put in another way, if the coordinates of a point (x, y) are multiplied by a (2×2) matrix, the resulting coordinates (x', y') are unique to the values of x, y and the matrix. If the linear mapping lies further from the origin than the initial point, the process is described as expansion. The reverse process is contraction.

The numbers forming the coordinates of a point are called scalars. When plotted on a graph, a scalar is represented by a position on one of the axes, for example the point $(2, 1)$ represents a scalar of two units on the x-axis and a scalar of 1 unit on the y-axis. The point within the area of the graph, defined by the two scalars, is called a vector, because when considered alone, it is assumed to represent a straight line, having a specific length and direction, joining the point to the origin. Thus for example, the column vector

$$\begin{bmatrix} 2 \\ 1 \end{bmatrix}$$

represents the line on a two-dimensional plot, joining the origin to the coordinates $(2, 1)$. It has direction, given by the ratio 2/1, and also magnitude, which is equal to $\sqrt{(2^2 + 1^2)} = \sqrt{5}$.

The third order column vector

$$\begin{bmatrix} x \\ y \\ z \end{bmatrix}$$

represents a line from the coordinates (x, y, z) in three-dimensional space, to the origin of a three-dimensional plot. With higher order vectors, there are not enough spatial dimensions to represent the vector, either on paper or as a model. The concept becomes imaginary, but can be represented in matrix form.

It has been explained above that if the coordinates of a point on a graph are multiplied by a second order matrix, the resulting pair of scalars is unique to the original coordinates and the matrix. Thus for a given (2×2) matrix, every possible position on a graph has a corresponding linear mapping, and the lines joining each point to its mapping will form a series of lines, scattered around the origin and orientated in a range of directions. It is known that for a given (2×2) matrix, there will only be two such lines which will extrapolate to the origin. The slopes of these lines are called the eigenvectors of the matrix.

It will be shown later that one of the eigenvectors of the matrix

$$\begin{bmatrix} 4 & 1 \\ 2 & 3 \end{bmatrix}$$

is unity, which means that for that specific matrix, any vector whose coordinates are equal will have a direction which is the same as that of its linear mapping. As an example, the column vector

$$\begin{bmatrix} 1 \\ 1 \end{bmatrix}$$

has coordinates which are equal, and therefore the slope is unity. Using the procedure shown in Fig. 6.1, the linear mapping (x', y') for the combination of the above vector and matrix is given by equation (6.20). The new vector

$$\begin{bmatrix} 5 \\ 5 \end{bmatrix}$$

$$\begin{bmatrix} x' \\ y' \end{bmatrix} = \begin{bmatrix} 4 & 1 \\ 2 & 3 \end{bmatrix} \begin{bmatrix} 1 \\ 1 \end{bmatrix} = \begin{matrix} 4 \times 1 + 1 \times 1 \\ 2 \times 1 + 3 \times 1 \end{matrix} = \begin{bmatrix} 5 \\ 5 \end{bmatrix} \tag{6.20}$$

obviously has the same direction as the vector

$$\begin{bmatrix} 1 \\ 1 \end{bmatrix},$$

as shown in Fig. 6.1. The other eigenvector of the matrix will be shown to be -0.5, hence the linear mapping of the matrix

$$\begin{bmatrix} 4 & 1 \\ 2 & 3 \end{bmatrix}$$

and the column vector,

$$\begin{bmatrix} 0.5 \\ -1.0 \end{bmatrix}$$

will be

$$\begin{bmatrix} 1 \\ -2 \end{bmatrix}$$

which has the same direction. Only when the coordinates of the original vector are in one of two ratios, 1.0 or -0.5 (the two eigenvectors), will the straight line joining the point to its linear mapping pass through the origin.

Thus in general terms, using the symbols used in equations (6.18) and (6.19),

$$\begin{bmatrix} x' \\ y' \end{bmatrix} = \begin{bmatrix} a & b \\ c & d \end{bmatrix} \begin{bmatrix} x \\ y \end{bmatrix} \tag{6.21}$$

where x and y and x' and y' are each in the same proportion as one of the eigenvectors. The (2×2) matrix can therefore be represented by a single constant λ which has two possible values, one for each eigenvector. These are the *eigenvalues*. Equation (6.21) then takes the form of equation (6.22)

$$\lambda \begin{bmatrix} x \\ y \end{bmatrix} = \begin{bmatrix} a & b \\ c & d \end{bmatrix} \begin{bmatrix} x \\ y \end{bmatrix} \tag{6.22}$$

Calculation of eigenvectors and eigenvalues, using MINITAB

$$\begin{bmatrix} 4 & 1 \\ 2 & 3 \end{bmatrix}$$

is shown on p. 175. This gave the values displayed below.

Eigenvalues	Eigenvectors	
5.000	0.707	0.707
2.000	0.447	-0.894

This is the usual format for displaying such information. Inspection reveals that the sum of the eigenvalues is equal to the sum of the leading diagonal of the (2×2) matrix, i.e. $5+2=7$ and $4+3=7$. Also the sum of the squares of each line of eigenvectors is equal to unity, i.e. $2 \times (0.707)^2 = 1$ and $(0.447)^2 + (-0.894)^2 = 1$. Each row represents a principal component. The first principal component is $0.707x + 0.707y$, and the eigenvalue represents the fraction of the total *variance* which is explained by that component. Thus the first principal component explains

$$\frac{5 \times 100}{5+2} = 71.4\%$$

of the variance.

CLUSTER ANALYSIS

Cluster analysis is used to classify results into groups, or clusters of closely related results. The simplest forms of cluster analysis are used for observations which are graded into two clearly distinct classes, for example active compounds and inactive compounds. The process is useful in assessing preliminary results, when quantitative tests have not been established, and the information obtained is only qualitative in nature. Since the results are not quantitative, regression analysis cannot be used. McFarland and Gans (1966) studied the monoamine oxidase (MAO) inhibiting properties of twenty aminotetralins and aminoindans, and found seven to be active. Table 6.11 summarizes the results. One-dimensional plots were then prepared, as

Table 6.11 — MAO activities and physicochemical parameters of some aminotetralins and aminoindans

Compound	Active (+) or inactive (−)	π	E_s	Random number
1	+	1.3	0.00	0.24
2	+	1.2	0.32	0.66
3	+	1.3	0.32	0.40
4	+	2.2	−0.07	0.17
5	+	1.7	0.00	0.58
6	+	1.0	0.00	0.08
7	+	0.8	0.32	0.66
8	−	1.7	0.00	0.46
9	−	1.7	−0.66	0.10
10	−	2.7	−0.66	0.98
11	−	4.2	−0.68	0.90
12	−	3.5	−0.68	0.21
13	−	1.0	−0.66	0.42
14	−	1.0	0.00	0.63
15	−	2.6	−1.08	0.76
16	−	2.6	−1.08	0.21
17	−	2.1	−1.08	0.54
18	−	0.8	0.32	0.65
19	−	1.4	−0.66	0.08
20	−	4.7	−0.68	0.96

shown in Fig. 6.3(a) to (c). The compounds were assigned random values in Fig. 6.3(a), Fig. 6.3(b) is scaled in terms of Hansch π values (Iwasa *et al.* 1965), which assess lipophilicity, and Fig. 6.3(c) uses Taft substituent parameters (E_s) Taft (1964), which are a measure of the bulk of the substituting group in the molecule. Scrutiny of the plots reveals that, with the exception of a few outliers, the inactive compounds are clustered towards the high π values and the low E_s values, suggesting that

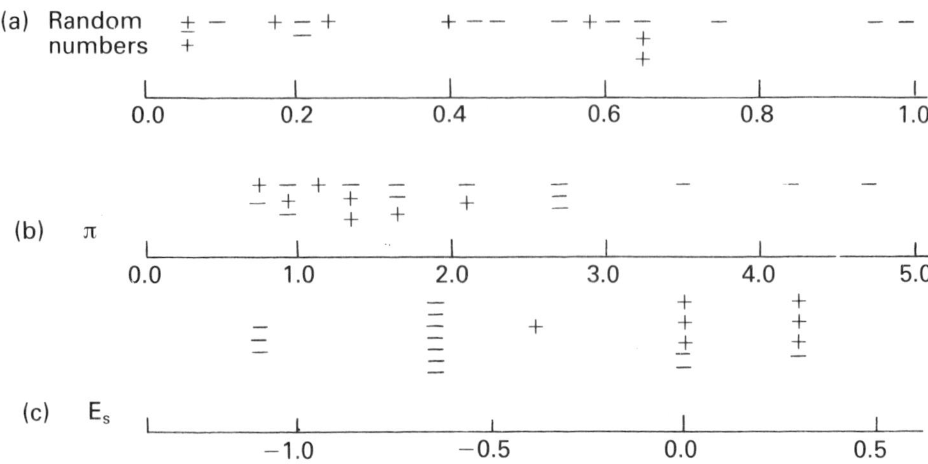

Fig. 6.3 — One-dimensional cluster plot of MAO activities and physiochemical parameters (+ active; − inactive).

biological activity is dependent on low lipid solubility and the absence of large substituent groups.

An extension is to prepare a two-dimensional plot, as shown in Fig. 6.4. The positions of the points confirm the dependence of MAO inhibition on steric factors, and the fact that all the active compounds have low π values confirms the importance of low lipid solubility. The procedures applied to these results are explained in more detail in McFarland and Gans's paper, together with their application to more complicated systems. It can be used equally well with quantitative data, by choosing an activity threshold, below which the observation is considered to represent inactivity. The precise value of the threshold is not critical, because decisions are based on recognition of patterns, which allows latitude with respect to the level at which the threshold is pitched.

Fig. 6.4 leaves little doubt regarding the positions of the two clusters with respect to π and E_s. However, sometimes there is a greater degree of overlapping of the points, so that allocation to clusters is not clear cut; it may even be questionable whether or not the total population can reasonably be resolved into more than one cluster. The problem can be resolved by determining the mean squared distance (MSD) between the points, as defined in equation (6.23) for the π values of a possible cluster involving the first three compounds in Table 6.11.

$$\mathrm{MSD} = \frac{(\pi_1 - \pi_2)^2 + (\pi_1 - \pi_3)^2 + (\pi_2 - \pi_3)^2}{3}$$

$$= \frac{(1.3 - 1.2)^2 + (1.3 - 1.3)^2 + (1.2 - 1.3)^2}{3} = 0.0067 \qquad (6.23)$$

The smaller the MSD, the more tightly bound the cluster, and the more probable that

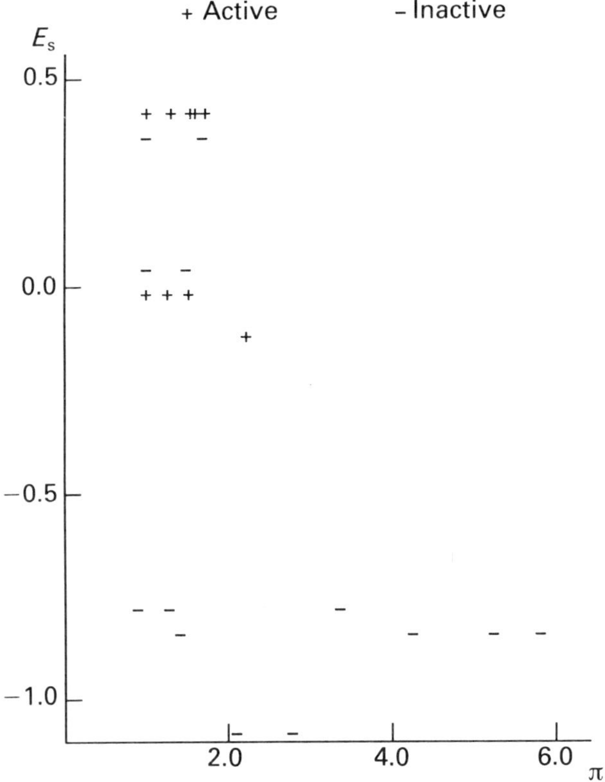

Fig. 6.4 — Two-dimensional plot of MAO activities and physicochemical parameters.

lipohilicity, as defined by the π values, is an important determinant of activity. To establish if the cluster is real, and not due to chance, MSD's must be calculated for all other combinations of three observations. Suppose for example, the compounds 1, 2 and 3 in Table 6.11, with an MSD of 0.0067, were part of a population of six compounds, the remainder being compounds 18, 19 and 20. There would be 20 possible combinations of three compounds, analogous to the expression in equation (6.23), and the probability (P) that another cluster of three compounds, which is more or equally condensed than the one comprising compounds 1, 2 and 3, will be given by equation (6.24).

$$P = A/20 \tag{6.24}$$

A is the number of groups of three compounds, including the group under test, that have MSDs equal to or less than the MSD of the test group. The number 20 represents the total number of combinations which are possible. For a large group of numbers, A can be estimated from a random sample taken from the total population.

The procedure can be applied to problems outside quantitative structure–activity relationships, for example solubilities in solvent blends, using solubility parameter or dielectric constant of the solvent mixtures as parameters.

A similar approach can be made with two-dimensional plots, for example with Fig. 6.4 the MSD of compounds 1 to 3 would be given by equation (6.25). However, for all 20 compounds there will be 77 520 combinations, taken 7 at a time, so that a computer would be required to make the necessary calculations.

$$\text{MSD} = \frac{(\pi_1 - \pi_2)^2 + (E_{s1} - E_{s2})^2 + (\pi_1 - \pi_3)^2 + (E_{s1} - E_{s3})^2}{3} + \frac{(\pi_2 - \pi_3)^2 + (E_{s2} - E_{s3})^2}{3}$$
$$= \frac{(1.3 - 1.2)^2 + (0 - 0.32)^2 + (1.3 - 1.3)^2 + (0 - 0.32)^2}{3} + \frac{(1.2 - 1.3)^2 + (0.32 - 0.32)^2}{3} = 0.0749 \quad (6.25)$$

DENDROGRAMS

Hierarchic or agglomerative methods

Earlier in this chapter (Table 6.1), five samples of olive oil were examined to determine which two samples were nearest with respect to five analytical properties. Cluster analysis is an extension of this, in which samples are classified into clusters of nearest neighbours. The procedure does not determine the number of groups, but given the number of groups required, it selects which samples go into which clusters. Considering Table 6.1, the samples initially fall into five groups, A, B, C, D and E. The distance matrix given in Table 6.3 tells us that C and E are closest in properties, so that if we wish to classify the data into four clusters, we would combine C and E, which are separated by only 1.581 units, to give (C,E),A, B and D. B and E are the next nearest to each other (2.251), so that for three clusters the arrangement is (B,C,E),A and D, and for two clusters it is (A,B,C,E) and D, because the next nearest neighbours are B and A, with a distance of separation of 2.561 units. This information can be plotted in the form of a dendrogram, in which the clusters are arranged along the abscissa and the distances between the clusters form the ordinate. The dendrogram can be plotted in terms of nearest neighbours, as in Fig. 6.5(a) or of the furthest neighbour distance, as shown in Fig. 6.5(b). This has the same overall shape as the nearest neighbour plot, but the heights of the blocks are greater. The distance between C and E is fixed at 1.581 units, and so this block has the same height in both plots, but the second cluster takes the greatest distance between a pair from B,C and E, which is 3.114 units. Similarly, the furthest distances for A,B,C,E and A,B,C,D,E are 3.915 and 4.309 resepctively.

It will be noticed that the samples are not arranged in alphabetical order in Fig. 6.5. This is because the samples should be arranged in a manner in which the information is most easily understood. Thus in the present situation, if A,B,C,D and

(a) Nearest neighbour plot

(b) Furthest neighbour plot

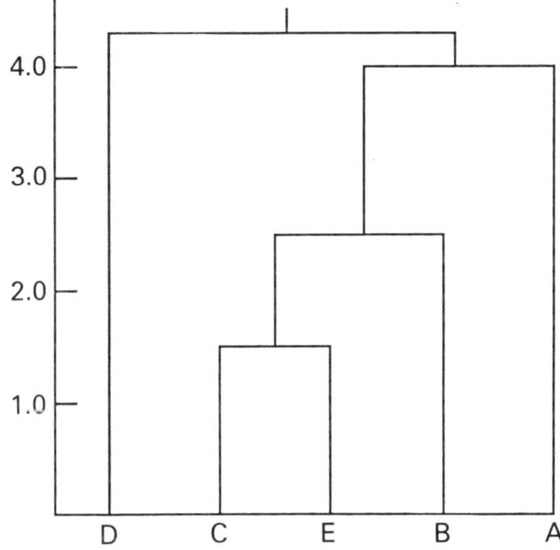

Fig. 6.5 — Dendrograms for olive oil samples. (For convenience, distances are to the nearest 0.25 units.)

E were arranged in alphabetical order, it would be impossible to plot a dendrogram. Wherever possible, it is best to place clusters with the smallest distances of separation furthest apart. If for example, the matrix shown in Table 6.3 took the form shown in Table 6.12, the nearest distance dendrogram would be preferentially presented in the form shown in Fig. 6.6, in which the two shortest distances (AB) and (CE) are placed furthest apart.

Table 6.12 — Rearranged distance matrix

	A	B	C	D	E
A	—				
B	2.25	—			
C	3.92	2.86	—		
D	3.99	2.56	4.31	—	
E	3.31	3.18	1.58	3.11	—

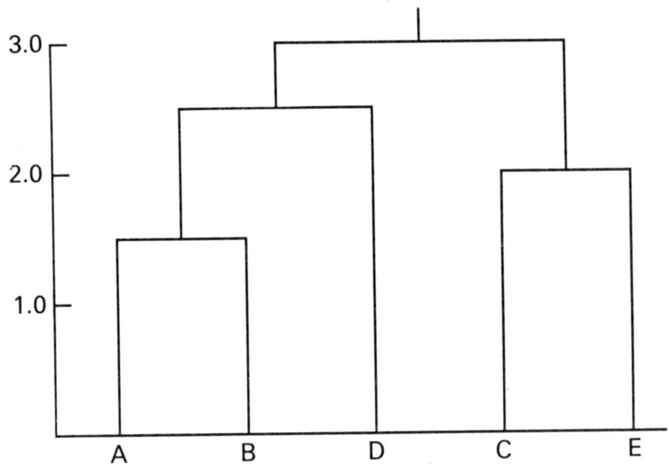

Fig. 6.6 — Alternative dendrogram for olive oil samples.

Another example can be taken from Table 6.13, which gives the results of an aerosol investigation (Evans and Farr) in which nine valves were compared with respect to the resulting 'respirable fraction', described as the fraction of the discharge containing globules greater than 50 μm in diameter. Three formulations were tested, giving 27 results. Standardized respirable fractions were obtained, using the mean and standard deviation of the 27 results. Thus for the first row of Table 6.13,

$$\text{Standardized value} = \frac{32.38 - 24.86}{7.53} = 0.999$$

Mean values for each valve where then calculated, for example for valve RD,

$$\text{Mean} = \frac{0.999 - 0.398 + 0.660}{3} = 0.420$$

Table 6.13 — Respirable fractions of three aerosol formulations with nine metered valves

Valve	Respirable fraction	
	Raw	Standardized
1% Surfactant, 4.28% mole fraction water		
1. RD	32.38	0.999
2. GR	18.76	−0.810
3. YE	17.25	−1.011
4. BK	24.02	−0.112
5. PK	24.47	−0.052
6. DB	30.47	0.745
7. YF	26.43	0.208
8. CR	41.26	2.178
9. LB	32.78	1.052
2% Surfactant, 2.82% mole fraction water		
10. RD	21.86	−0.398
11. GR	14.51	−1.375
12. YE	13.69	−1.483
13. BK	18.84	−0.799
14. PK	15.04	−1.304
15. DB	21.92	−0.390
16. YF	22.18	−0.356
17. CR	33.09	1.093
18. LB	19.11	−0.764
2% Surfactant, 1.42% mole fraction water		
19. RD	29.83	0.660
20. GR	20.36	−0.598
21. YE	15.57	−1.234
22. BK	24.62	−0.032
23. PK	21.04	−0.507
24. DB	29.45	0.610
25. YF	29.83	0.660
26. CR	39.07	1.887
27. LB	33.47	1.143
Mean of 27 results	24.86	
Standard deviation	7.53	

Table 6.14 gives the complete set of means, from which the distance matrix, also shown in Table 6.14, and the dendrogram shown in Fig. 6.7 were prepared. The results clearly separate into two clusters, DB, RD, LB, YF, GR and PK, BK, YE.

Table 6.14 — Mean standardized respirable fractions and distance matrix for 9 metered aerosol valves

Valve	RD	GR	YE	BK	PK	DB	YF	CR	LB
mean	0.420	−0.928	−1.243	−0.314	−0.621	1.031	0.512	1.719	0.477
		1.438	1.664	0.734	1.041	0.611	0.092	1.299	0.050
			0.315	0.614	0.307	1.959	1.440	2.647	1.405
				0.929	0.622	2.274	1.755	2.962	1.720
					0.307	1.345	0.826	2.033	0.791
						1.652	1.133	2.340	1.098
							0.519	0.688	0.554
								1.207	0.035
									1.242

Valve CR appears to form a cluster of its own, and at the same time gives the highest respirable fraction.

Dendrograms can also be constructed using correlation coefficients, when one is concerned with relationships between variables, rather than similarities. Table 6.15 can be used as an example. It gives a list of questions given to a panel of ladies in a comparison of three shampoo formulations (Harris *et al.* 1975; Powell *et al.* 1973). Assessments were recorded on a six-inch line, one end of which was labelled 'very poor', and the other 'very good'. Subjects were asked to draw a stroke across the line, and the judgement was assessed by measuring the distance, in inches, between the 'very good' end and the crossing point, so that the smaller the distance, the better the rating. A correlation matrix of the scalings for one of the formulations is given in Table 6.16, and the dendrogram drawn in Fig. 6.8. Questions 12 to 15 received answers which were obviously not correlated with any of the remainder, and so were omitted from the dendrogram. Questions 1 and 2 were the most highly correlated, because the height of the rectangle joining 1 and 2 is lower than all the others. This is not surprising, because the conclusion drawn from question 2 would form the basis of the answer to question 1. Questions 9 and 12 are also highly correlated, indicating that 'manageable' and 'tangle free' are synonymous to the subjects, as also are 'feel' and 'texture' (questions 10 and 11). Evaluations involving lather and cleansing (questions 3, 4 and 5) are clustered together and associated, not surprisingly, with question 16, 'Did you have enough shampoo?' Another cluster (7,6,1,2 and 8) suggests that ladies like shampoos which leave their hair in good condition, which they judge in terms of 'shine' and 'body'.

Partitioning methods

In these methods, results are allocated arbitrarily to groups, and the populations of the groups then adjusted to fit the needs of the exercise. The technique is used in planning quantitative structure–activity relationship (QSAR) experiments. QSAR is a study in which the biological activities of a collection of related compounds are used

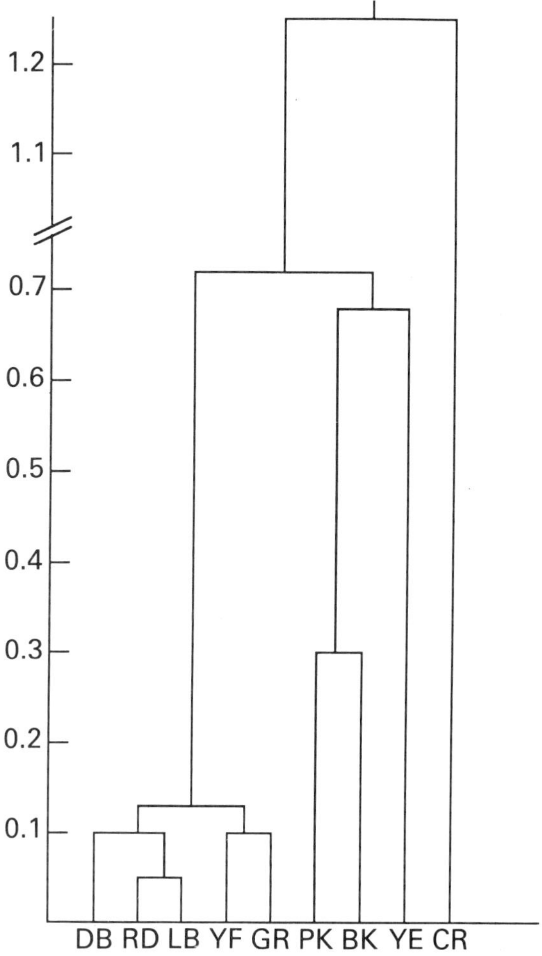

Fig. 6.7 — Cluster analysis for aerosol valves.

to predict what new derivatives are likely to show promise. The procedure depends on multiple regression analysis of a biological parameter against physicochemical parameters, such as the Hammett substituent constant (σ) (Hammett 1940), which is an indicator of electron density, or the Hansch constant (π) (Iwasa *et al.* 1965), which is related to octanol–water partition coefficients, and is a measure of the lipophilic influence of the substituent groups. QSAR is reviewed elsewhere, for example (James 1974, 1988, Hansch and Leo 1979). In multiple regression analysis, it is important that none of the so-called independent variables are inter-related. Such relationships have been demonstrated in QSAR, and some combinations of chemical groups exhibit more colinearity than other combinations. An obvious example is the relationship between partition coefficients and molar volumes of homologous series, since both parameters increase uniformly with each additional methylene group.

Table 6.15 — Questions asked in shampoo trial

1. Did you like the shampoo?
2. Did you think the shampoo was suitable for your hair type?
3. What did you think of the lathering ability of the shampoo?
4. What did you think of the 'rinsibility' of the shampoo?
5. What did you think of the shampoo's cleansing power?
6. How would you judge the shampoo for the condition it left your hair?
7. How would you judge the shampoo for the shine it gave your hair?
8. How would you judge the shampoo for the 'body' it gave your hair?
9. How would you judge the shampoo for how manageable it left your hair?
10. How would you judge the shampoo for the feel it gave your hair?
11. How would you judge the shampoo for the texture it gave your hair?
12. How would you judge the shampoo for how 'tangle free' it left your hair?
13. How would you judge the shampoo for how oily it left your hair?
14. How would you judge the shampoo for how mild it was to your skin?
15. How would you judge the shampoo for its perfume?
16. Did you have enough shampoo?

Table 6.16 — Correlation matrix between questions asked in shampoo trial

Question number	1	2	3	4	5	6	7	8	9	10	11	12	13	14	15	16
1		.86	.47	.33	.60	.76	.45	.62	.34	.41	.27	.36	.11	.22	.09	.23
2			.35	.33	.49	.72	.36	.66	.33	.31	.21	.31	.10	.17	.08	.14
3				.47	.63	.36	.35	.18	.15	.39	.22	.28	.09	.40	.15	.55
4					.59	.29	.11	.39	.46	.28	.19	.45	.04	.41	.02	.23
5						.60	.45	.45	.41	.54	.35	.42	.03	.38	.03	.29
6							.54	.62	.41	.43	.32	.41	.03	.08	.07	.11
7								.33	.26	.35	.34	.28	.21	.00	.08	.11
8									.45	.32	.16	.37	.01	.26	.05	.04
9										.29	.32	.81	.27	.27	.15	.10
10											.72	.32	.01	.32	.16	.20
11												.41	.03	.20	.02	.08
12													.25	.25	.21	.00
13														.04	.01	.11
14															.26	.04
15																.12
16																

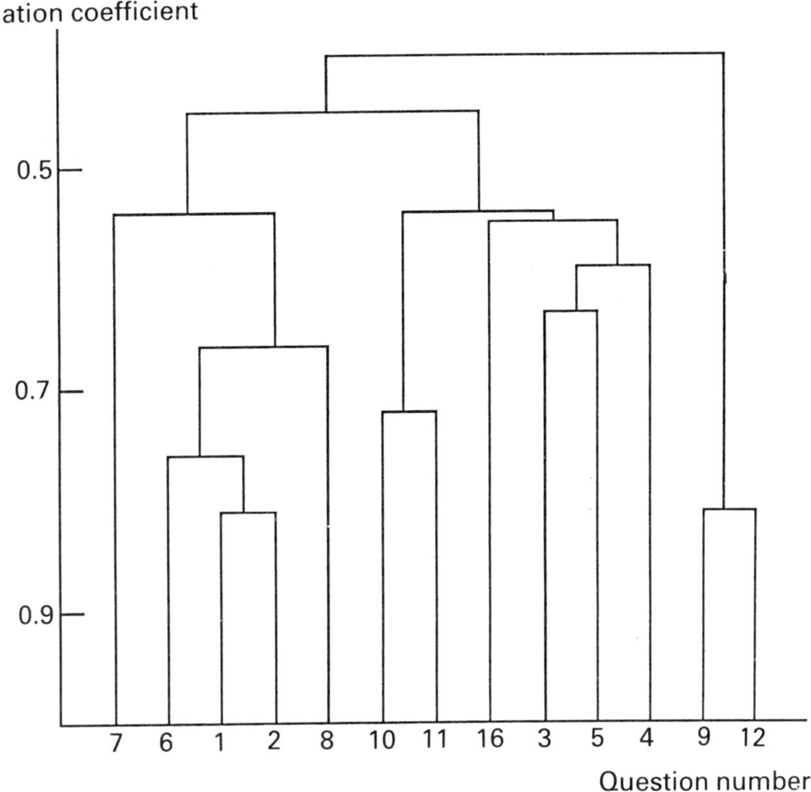

Fig. 6.8 — Dendrogram of shampoo evaluation results.

Hansch *et al.* (1973) devised a scheme for avoiding such problems. They considered six parameters π and π^2, which are measures of lipophilicity, Swain and Lupton's (1968) inductive and resonance constants ($\mathcal{F}$ and $\mathcal{R}$), which are related to electron density, and molar refractivity (MR) and molecular weight (MW) which are related to molecular bulk. They studied 27 substituent groups in 166 compounds. The results were submitted to hierarchical clustering in six-dimensional space, using a procedure similar to that used to produce a dendrogram of the analytical parameters of the olive oil samples shown in Table 6.1. They forced the data into sets of 5, then 10 and then 20 clusters, so that for a QSAR exercise in which co-linearity was to be avoided, no two substituent groups should be selected from the same cluster. If there were an unlimited range of compounds available, the 20-cluster set was recommended, but the smaller cluster sets could be used when the number of compounds was restricted, with a corresponding risk of co-linearity. A fuller account of this procedure, together with the cluster sets can be seen elsewhere (Hansch and Leo 1979).

A simple example in two dimensions can be taken from Table 6.17, in which the Swain and Lupton inductive and resonance parameters of three halogens and three alkyl groups are given. The answers to the exercise are obvious, but the example serves to illustrate the procedure. Standardized values ($\mathcal{F}'$ and $\mathcal{R}'$) are obtained by subtracting the mean from each term and dividing by the standard deviation. Standardized values are given in Table 6.17. The mean squared distance (MSD)

Table 6.17 — Swain and Lupton parameters for some alkyl and halogen groups

	CH_3	C_2H_5	C_3H_7	Cl	Br	I	Mean	St.Dev.
$\mathcal{F}$	−0.04	−0.05	−0.06	0.41	0.44	0.40	0.183	0.256
$\mathcal{R}$	−0.13	−0.10	−0.08	−0.15	−0.17	−0.19	−0.137	0.042
Standardized values								
$\mathcal{F}$	−0.872	−0.911	−0.950	0.885	1.003	0.846		
$\mathcal{R}$	0.160	0.877	1.356	−0.319	−0.798	−1.276		

between CH_3 and C_2H_5, for example, is then given by equation (6.26). The remaining values are shown in Table 6.18.

Table 6.18 — Distance matrix of electronic parameters from Table 6.17

	CH_3	C_2H_5	C_3H_7	Cl	Br	I
CH_3		0.52	1.44	3.32	4.43	5.01
C_2H_5			0.23	4.66	6.47	7.72
C_3H_7				6.15	8.45	10.15
Cl					0.24	0.92
Br						0.25
I						

$$dCH_3,C_2H_5 = (-0.872 + 0.911)^2 + (0.160 - 0.877)^2 = 0.52 \qquad (6.26)$$

The corresponding dendrogram is shown in Fig. 6.9, in which the alkyl groups and the halogens form two distinct clusters. The indication is therefore that when the inductive and resonance constants $\mathcal{F}$ and $\mathcal{R}$ are used in a QSAR exercise, no more than one representative should be taken from either cluster.

DISCRIMINATION

Discrimination analysis can be regarded, very loosely, as the reverse of cluster analysis. In cluster analysis the data are processed as a whole, with the object of identifying groups of related results, while in discrimination the data are initially

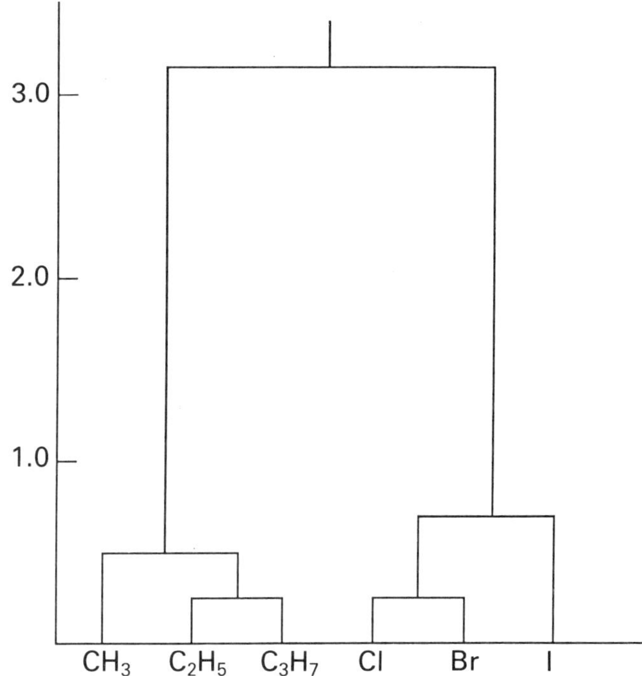

Fig. 6.9 — Dendrogram of electronic parameters for halogens and lower alkyl groups.

divided into groups, according to a preconceived hypothesis, and the credibility of the classification then assessed. In the simplest situation, the hypothesis that a collection of values of one variable is divisible into two subgroups can be tested by plotting the variable on a scatter diagram, in the same way as was employed with clustering in Fig. 6.4. The hypothesis can be tested by visual observation, and is characterized by the points separating into two groups. A subsequent, more sophisticated treatment could use a test for significance, as with Student's *t* test (Page 11), from which the probability of there being two groups can be assessed.

A similar exercise is possible with two variables, as shown in Fig. 6.10. The variables may be directly related, giving one or two straight lines (Fig. 6.10(a) and (b)), or partially correlated, giving eliptical plots (Fig. 6.10(c) and (d)). The existence of one or two groups can be judged from the degree or absence of overlap of points, which have been tentatively allocated to different groups. Alternatively, the two variables may be independent, giving scattered plots (Fig. 6.10(e) and (f)), but also providing a subjective means of discriminating between two groups.

Scatter diagrams can therefore be used to establish the existence of two or more subgroups within the complete data set. It can also be used to allocate new results to their respective sets by ascertaining where they lie in the diagram. The process has the advantage that the number of individual results within each group need not be the same. However, it is essential that the variances within the groups are similar,

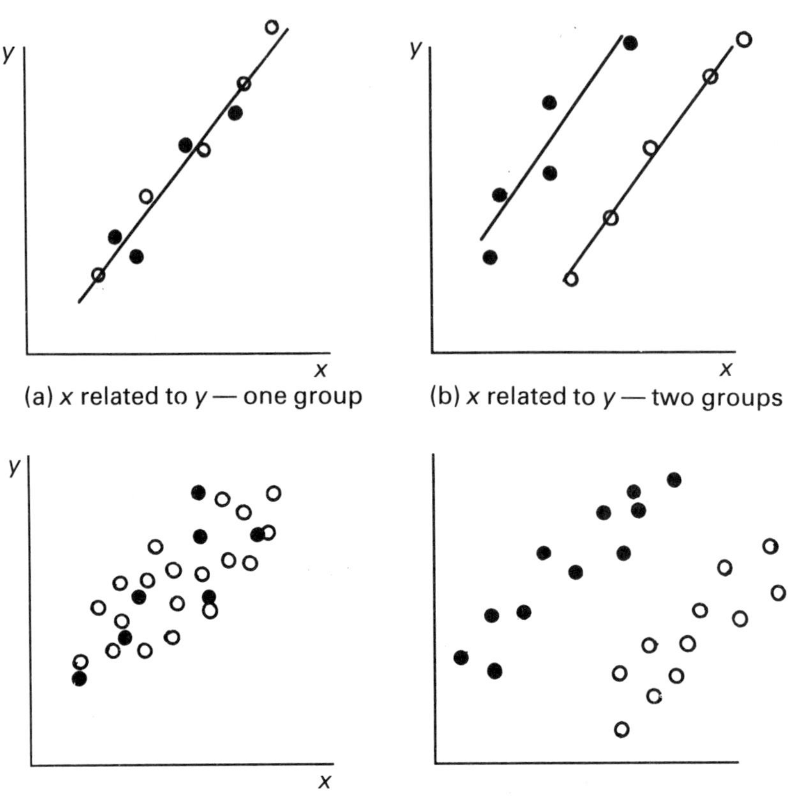

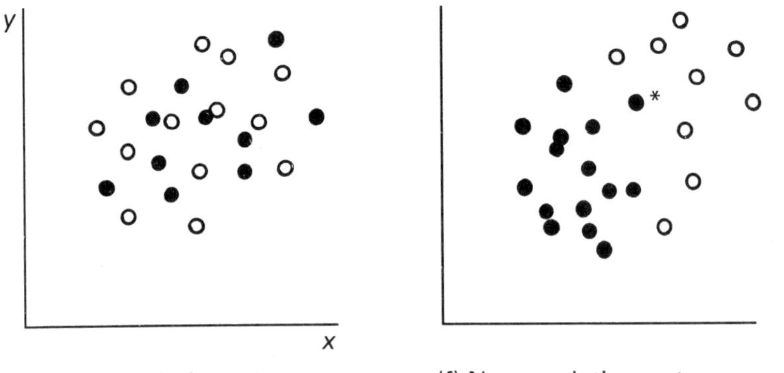

Fig. 6.10 — Scatter diagrams with and without discrimination.

otherwise the outcome would be biased in favour of the variables with the greatest variance. For this reason, is it advisable to standardize the data.

When assignment to a group is ambiguous, it becomes necessary to calculate to which cluster an individual result belongs. Such calculations are dependent on the distance between the point and a position representative of the profile, usually the mean coordinates. Thus in two dimensions, the distance (d) between a point (x, y) and the average value of cluster A (x_A, y_A), is given by equation (6.27). Similarly, the distance from the average of cluster B is given by equation (6.20), so that if $d_A < d_B$, the point belongs to cluster A, and *vice versa*.

$$d_A = \sqrt{(x - x_A)^2 + (y - y_A)^2} \tag{6.27}$$

$$d_B = \sqrt{(x - x_B)^2 + (y - y_B)^2} \tag{6.28}$$

The method can be extended to any number of dimensions, and can be demonstrated using the results given in Table 6.19 (Rushton, 1988), which classifies the scalp hair of ten males and seven females according to three parameters:

(1) The total number of hairs per square centimetre.
(2) The number of hairs per square centimetre of diameter greater than 40 μm and length greater then 30 mm, these dimensions being the lower limit of fibres which contribute to the aesthetic quality of the hair.
(3) The number of actively growing hairs per square centimetre.

The object of the exercise is to determine if male hair follows a different profile from female hair. Scores in each column are standardized by substracting the means for males plus females, and dividing by the corresponding standard deviation. Thus for example, for the total number of hairs on the first male in Table 6.19,

$$\text{Standardized result} = \frac{177 - 210.2}{45} = -0.74$$

Standardized means are given in Table 6.19. The root mean square distance between each standardized result and the male and female standardized means are then calculated, for example for the first male, comparison with the average male gives

$$d_A = \sqrt{(-0.74 - 0.277)^2 + (-0.60 - 0.398)^2 + (-1.05 + 0.439)^2}$$
$$= 1.55$$

Similarly, comparison with the average female gives,

$$d_B = \sqrt{(-0.74 + 0.396)^2 + (-0.60 + 0.570)^2 + (-1.05 - 0.623)^2}$$
$$= 1.71$$

Table 6.19 — Parameters for male and female hair

	Total hairs		Hairs >40 μm and >30 mm		Growing hair	
	No.	Standardized	No.	Standardized	No.	Standardized
Males						
1	177	−0.74	154	−0.60	61.9	−1.05
2	175	−0.78	159	−0.48	48.6	−2.21
3	180	−0.67	167	−0.29	59.8	−1.23
4	231	0.46	219	0.97	73.4	−0.05
5	222	0.26	193	0.34	75.2	0.10
6	218	0.17	174	−0.12	61.5	−1.09
7	240	0.66	208	0.70	79.5	0.48
8	239	0.64	205	0.63	72.8	−0.10
9	276	1.46	234	1.33	67.4	−0.57
10	269	1.31	241	1.50	89.3	1.33
Mean	222.7	0.277*	195.4	0.398	68.9	−0.439*
Standard deviation	36.3		31.1		11.6	
Females						
1	125	−1.89	108	−1.71	86.8	1.11
2	154	−1.25	138	−0.99	80.1	0.53
3	156	−1.20	137	−1.01	75.2	0.10
4	172	−0.85	113	−1.59	75.1	0.10
5	236	0.57	223	1.06	88.0	1.22
6	259	1.08	158	−0.50	73.1	−0.08
7	245	0.77	210	0.75	89.9	1.38
Mean	192.4	−0.396*	155.3	−0.570*	81.2	0.623*
Standard deviation	53.0		45.2		6.99	
Statistical parameters for male plus female						
Mean	210.2		178.9	−0.569	74.0	
Standard deviation	45.0		41.5		11.5	

* These means have finite values, rather than zero, because the standardized results were calculated using the means and standard deviations for all 17 subjects.

1.55 is less than 1.71, so male 1 is assigned to the male group. The complete results are presented in Table 6.20. In all cases, the distances between male subjects and the average male are less then the distances from the average female. For similar reasons, the females fall into a separate group. There is one exception, female 6 gives virtually the same Euclidean distance from both male and female means; this must be considered to be an interface result, as exemplified by the point marked with an asterisk in Fig. 6.10(f). It may be concluded that, according to the three parameters

Table 6.20 — Euclidean distances between male and female hair parameters

	Male		Female	
	Versus male	Versus female	Versus male	Versus female
1	1.55	1.71	3.40	1.94
2	2.24	2.86	2.28	0.96
3	1.41	1.89	2.11	1.06
4	0.72	1.89	2.35	1.23
5	0.54	1.24	1.81	1.99
6	0.84	1.86	1.26	1.64
7	1.04	1.66	1.92	1.92
8	0.55	1.74	—	—
9	1.51	2.91	—	—
10	2.33	2.77	—	—

quoted, male hair is different from female hair, and therefore discrimination into two groups is justified.

A similar problem is given in Table 6.21, which shows the respirable fractions

Table 6.21 — Discriminant analysis of aerosol formulations

	Standardized respirable fractions			Euclidian distances		
Formula / Valve	A	B	C	$\sqrt{(x-x_A)^2}$	$\sqrt{(x-x_B)^2}$	$\sqrt{(x-x_C)^2}$
RD	0.999	−0.398	0.660	0.644	1.641	0.711
GR	−0.810	−1.375	−0.598	1.166	0.168	1.098
YE	−1.011	−1.483	−1.234	1.366	0.369	1.299
BK	−0.112	−0.799	−0.032	0.467	0.530	0.400
PK	−0.052	−1.304	−0.507	0.407	0.590	0.340
DB	0.745	−0.390	0.610	0.390	1.387	0.457
YF	0.208	−0.356	0.660	0.147	0.850	0.080
CR	2.178	1.093	1.887	1.823	2.820	1.890
LB	1.052	−0.764	1.143	0.697	1.694	0.764
$\bar{x}$	0.355	−0.642	0.288			

obtained from three aerosol formulations using nine valves (Evans and Furr n.d.). It has already been established by cluster analysis (page 102) that the valves can be classified into three sets, and it is now required to know if the respirable fractions fall into three groups, one for each formulation. Standardized respirable fractions are given in Table 6.13. These are grouped into the three sets in Table 6.21, together with the mean standardized respirable fractions for each set. Euclidean distances between each standardized result and each set mean are given in Table 6.21. Thus for valve *RD* and formula *A*,

$$\sqrt{(x-\bar{x}_A)^2} = \sqrt{(0.999-0.355)^2} = 0.634; \qquad \sqrt{(x-\bar{x}_B)^2} =$$

$$\sqrt{(0.999+0.642)^2} = 1.641; \qquad \sqrt{(x-\bar{x}_C)^2} = \sqrt{(0.999-0.181)^2} = 0818$$

The first result is the smallest, so with the RD valve, formula *A* falls into cluster *A*. However, Table 6.21 shows that this is not always the case, for example formula *A* with valve *GR* appears to belong to cluster *B*, while formula *A* with valve *BK* falls into cluster *C*. It is therefore evident that the results cannot be classified with respect to formulations, and there is no point in continuing the calculation further.

A disadvantage with the above procedure is that no account is taken of correlation between variables. When variables are related they are, at least in part, measuring the same thing, and yet they carry the same weight as the independent variables. Equation (6.29) defines the Mahalanobis distance (D_m), which takes correlations into consideration. *A* and *B* represent two postulated clusters, and V^{-1} is derived from the covariance matrix. The *S* symbols signify the sums of the bracketed terms with respect to variables. The Mahalonobis method is more protracted than the one described above, and requires an accurate covariance matrix.

$$D_m^2 = S(x_A - \bar{x}_A)^{V^{-1}} S(x_B - \bar{x}_B) \tag{6.29}$$

PRINCIPAL COMPONENTS ANALYSIS

The object of principal components analysis is to reduce the number of variables necessary to characterize an array of numbers. Thus if there are *p* columns, C_1, C_2, $C_3 \ldots C_p$, the analysis seeks to find combinations of C_1 to C_p, so that less than *p* such combinations are necessary to characterize the system. The viscosities of glycerol/water mixtures given in Table 5.1 can be used as an example. Standardization of these results give the (2×5) matrix shown in Table 6.22(a), and yields the covariance matrix shown in Table 6.22(b). Eigenvalues and eigenvectors are given in Table 6.22(c).

The methodology for calculating these parameters is given elsewhere, as follows, standardized data, page 83, correlation matrix, page 86 and eigenvalues and eigenvectors, page 91.

The following observations and inferences can be made from the information in Table 6.22.

Table 6.22 — Standardized results and correlation matrix of glycerol/water mixtures

(a) Standardized data	−1.267	−1.152	
	−0.629	−0.618	
	−0.002	−0.194	
	0.636	0.581	
	1.263	1.384	
(b) Correlation matrix	1.000	0.991	
	0.991	1.000	
(c) Eigenvalues and eigenvectors	Eigenvalues	Eigenvectors	
		% Glycerol	Viscosity
	1.992	0.707	0.707
	0.008	−0.707	0.707

(a) Both eigenvalues are positive. Negative values are not possible with a correlation matrix.
(b) The sum of the eigenvalues is equal to the sum of the leading diagonals in the correlation matrix, i.e. $1.992 + 0.008 = 1.000 + 1.000$.
(c) The sum of the squares of the eigenvectors in each row is equal to unity, i.e. $0.707^2 + 0.707^2 = 1.000$. This is termed the communality of the row. The importance of an eigenvector in a row can be assessed by calculating the communality without that vector, and noting how far it deviates from unity. This device is used later in the section on factor analysis.
These three conditions can be used to check the calculations.
(d) $(1.992 \times 100)/(1.992 + 0.008) = 99.6\%$ of the variance is explained by the first principal component. Only 0.4% of the variance is explained by the second principal component. This suggests that that the results can be defined in terms of only one variable, either the percentage glycerol or the viscosity, because one is related to the other. This of course has been demonstrated in Chapter 5, where it was shown that the two properties were rectilinearly related, with a correlation coefficient of 0.991. Regression analysis of the standardized results, given in Table 6.22, yields equation (6.30), which has a slope of approximately 1, and no intercept.

$$\text{Viscosity} = 0.991\ (\%\text{glycerol})\ \frac{1}{5}\ \frac{\text{r}}{0.992} \qquad (6.30)$$

(e) The two properties contribute equally to the 99.6% of the total variance explained by the first principal component.

It is obvious that using principal components analysis in this context is an excessively elaborate way of demonstrating what can be detected rapidly by linear regression analysis. However, the example serves to demonstrate in a simple fashion the functions of the technique. Relationships become less evident as the order of the matrix increases. Furthermore, correlation and covariance matrices only show relationships between pairs of variables. Relationships between three or more variables, encountered in higher order matrices, are less obvious, and it is under these circumstances that principal components analysis comes into its own.

The androgenic activities of testosterone esters, shown in Table 6.4, can be used to illustrate this. The correlation matrix of the standardized data, given in Table 6.6, shows a good correlation ($r = 0.995$) between log OAR and log k_c, but relationships between log OAR and either R_m or E_s are less definite, having correlation coefficients of less than 0.95 for only five sets of results. The determinants of the matrix are shown in Table 6.10, and confirm that relationships between pairs of results from log OAR, R_m and E_s are, at the best weak. Thus the determinant of the second order matrix of log OAR and log k_c is 0.010, while those for all other combinations are more than ten times greater. This is supported by the third order determinants, which are less than 0.0023 when both log OAR and log k_c are involved, but are about 10 times greater when only one of log OAR and log k_c is included. The low determinant (0.004) of the fourth order matrix is in line with these conclusions.

Eigenvalues and eigenvectors are given in Table 6.23. The first two eigenvalues

Table 6.23 — Eigenvalues and eigenvectors to Table 6.7

Eigenvalues	Eigenvectors			
3.73720	0.514221	0.512067	−0.494176	−0.478701
0.20961	−0.070132	−0.149625	0.559044	−0.812505
0.04929	−0.417714	−0.526439	−0.665460	−0.324869
0.00390	0.745770	−0.662014	0.020583	0.071702

are considerably greater than the others, and explain $(3.73720 + 0.20961) \times 100/4 =$ 98.6% of the variation in the results. The relationship between log OAR and log k_c obviously accounts for the one low eigenvalue, and the other probably reflects a relationship between log OAR and R_m. However, what is not disclosed is whether the relationship is a direct one between log OAR and R_m ($r = 0.944$) or is a consequence of a relationship between log k_c and R_m ($r = 0.946$).

To clarify this situation, it is necessary to consider the origins of the parameter log OAR. This was derived from the plot of organ weight (seminal vesicles plus prostate) against time after administration. A typical example is shown in Fig. 6.11. Overall androgen response (OAR) is the area under the plot, which is approximately triangular in shape, and is therefore equal to half the product of the height of the maximum point on the graph (BR_{max}) and the duration of androgenic activity, which in turn is proportional to the time at which the maximum response occurs, the time of

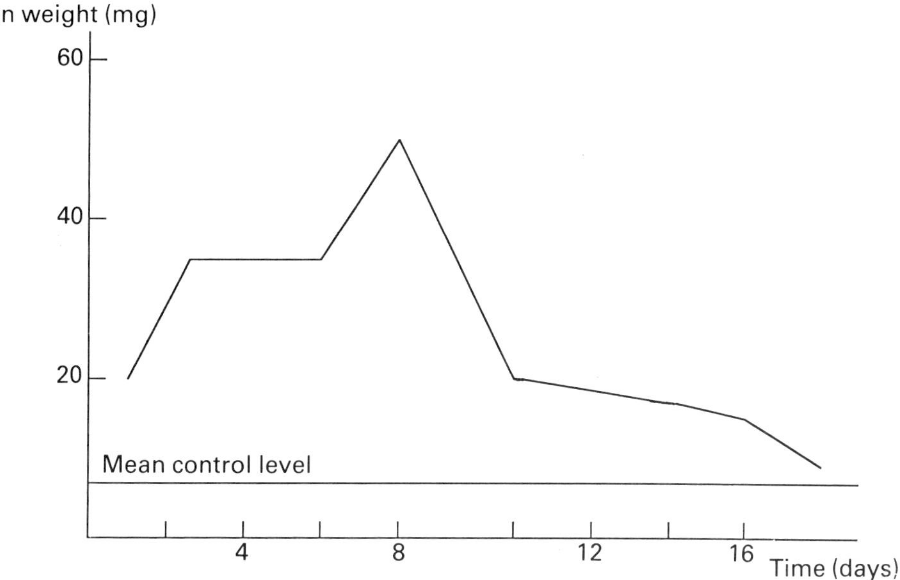

Fig. 6.11 — Increase in organ weight after a single injection of 200 μg of testosterone propionate.

maximum effect (TM). These additional parameters for the data in Table 6.4, are given in Table 6.24, and as shown in equation (6.31), are related in the suggested manner to log OAR.

Table 6.24 — Androgenic activities of some testosterone esters

Log TM	0.301	0.477	0.778	0.778	0.903
log BR_{max}	1.602	1.732	2.103	2.310	2.021

$$\log \mathrm{OAR} = -0.452 + 1.11(\log \mathrm{BR}_{max} + \log \mathrm{TM}) \quad \begin{matrix} n & r \\ 5 & 0.997 \end{matrix} \qquad (6.31)$$

This is confirmed in Table 6.25, in which the determinants of the second order matrices have finite values, but the determinant for the third order matrix is zero. Thus the three parameters are related, but no two of them are individually related. This information is not obvious from the covariance matrix, shown in Table 6.26, but it does detect relationships of both log k_c and R_m with TM, a relationship between log k_c and log BR_{max}, and another between log k_c and R_m. Eigenvalues and eigenvectors are given in Table 6.27, and show that a matrix consisting of five parameters can be

Table 6.25 — Determinants to log OAR, Log BR_{max} and log TM

Log OAR–log TM

$$\begin{vmatrix} 1.00000 & 0.96758 \\ 0.96758 & 1.00000 \end{vmatrix} = 0.064$$

$$\begin{vmatrix} 1.00000 & 0.96758 \\ 0.95341 & 0.85244 \end{vmatrix} = -0.070$$

$$\begin{vmatrix} 0.96758 & 1.00000 \\ 0.95341 & 0.85244 \end{vmatrix} = -0.129$$

log TM–log BR_{max}

$$\begin{vmatrix} 1.00000 & 0.85244 \\ 0.85244 & 1.00000 \end{vmatrix} = 0.273$$

$$\begin{vmatrix} 0.96758 & 0.95341 \\ 1.00000 & 0.85244 \end{vmatrix} = -0.129$$

$$\begin{vmatrix} 0.96758 & 0.95341 \\ 0.85244 & 1.00000 \end{vmatrix} = 0.155$$

log OAR–log BR_{max}

$$\begin{vmatrix} 1.00000 & 0.95341 \\ 0.96758 & 0.85244 \end{vmatrix} = -0.070$$

$$\begin{vmatrix} 0.96758 & 0.85244 \\ 0.95341 & 1.00000 \end{vmatrix} = 0.155$$

log OAR–log BR_{max}

$$\begin{vmatrix} 1.00000 & 0.95341 \\ 0.95341 & 1.00000 \end{vmatrix} = 0.091$$

log OAR-log TM-log BR_{max}

$$\begin{vmatrix} 1.00000 & 0.96758 & 0.95341 \\ 0.96758 & 1.00000 & 0.85244 \\ 0.95341 & 0.85244 & 1.00000 \end{vmatrix} = 0.00004$$

Table 6.26 — Covariance matrix of standardized values linking log TM and log BR_{max} with QSAR parameters

	log TM	log BR_{max}	log k_c	R_m	E_s
log TM	1.00000	0.85244	0.97679	−0.96602	−0.89614
log BR_{max}		1.00000	0.94316	−0.83248	−0.80773
log k_c			1.00000	−0.94602	0.88236
R_m				1.00000	0.79953
E_s					1.00000

Table 6.27 — Eigenvalues and eigenvectors to Table 6.26

Eigenvalues	Eigenvectors				
4.56548 (91.3%)	−0.460045	0.434437	0.465457	−0.445821	−0.429204
0.21299 (4.3%)	−0.028173	−0.209351	−0.143861	0.443681	−0.858970
0.20034 (4.1%)	0.369588	−0.800386	−0.095387	−0.460617	−0.038995
0.02118 (0.4%)	−0.541758	0.175651	−0.468704	−0.62025	−0.266920
0.00001 (0.0%)	−0.597886	−0.309783	0.730665	−0.086652	−0.072020

resolved into only three principal components, which will account for 99.5% of the variance in the results.

FACTOR ANALYSIS

This branch of multivariate analysis originated in a paper published by Spearman in 1904. The type of information upon which the paper was based is familiar to all who have been concerned with assembling the results of a multisubject examination. Table 6.28, which gives the marks awarded to a random sample of candidates presenting for a degree in pharmacy, is a typical example. The normal procedure is to average the number along each row, to give the overall performances of the candidates, and by placing these scores in numerical order, a rank order of achievement is obtained. In a similar way, the standards in each subject can be compared by calculating the means of the columns. This additional information is given in Table 6.28.

Spearman considered that a candidate's score in a particular subject depended on three factors, the candidate's general intelligence, his ability in the subject and the degree of difficulty of the subject. These are expressed mathematically in the form of equation (6.32).

$$x = aF + e \tag{6.32}$$

x is the score obtained by a given candidate in the given subject, F is a constant, specific to the subject and independent of the candidate, a is a constant, specific to the candidate, and e is a constant, the factor loading, which is related to both the subject and the candidate. Thus Table 6.28 gives 20 scores in pharmacology, which are related to 41 independent variables, 20 values of a, 20 values of e, and 1 value of F. There are therefore no unique values of a, F and e which satisfy equation (6.32) and the 20 pharmacology scores. Instead, an infinite range of values is possible; for

Table 6.28 — Examination scores for pharmacy degree candidates

Candidate	Subject score (%)						Mean score	Class position
	A	*B*	*C*	*D*	*E*	*F*		
1	30	51	44	38	35	37	39.2	20
2	48	43	61	52	58	50	52.0	12
3	52	54	72	68	59	51	59.3	6
4	41	46	56	65	56	24	48.0	15
5	52	62	65	46	61	57	57.2	9
6	56	67	72	73	51	49	61.3	4
7	51	43	58	42	62	57	52.2	11
8	41	40	51	53	54	57	49.3	14
9	48	68	59	58	56	55	57.3	8
10	56	87	70	73	65	66	69.5	1
11	44	63	57	43	52	46	50.8	13
12	51	69	62	71	60	44	59.5	5
13	58	69	75	63	68	67	66.7	2
14	48	56	45	47	51	39	47.7	16
15	57	71	70	71	65	63	66.2	3
16	53	58	55	56	52	40	52.3	10
17	42	48	54	42	54	37	46.2	17
18	35	43	50	39	50	41	43.0	19
19	36	59	47	50	39	42	45.5	18
20	45	65	66	66	55	48	57.5	7
Mean	47.20	58.10	59.45	55.80	55.15	48.50		
Standard deviation	7.84	12.22	9.44	12.19	8.07	10.95		

A = Pharmacology; *B* = Pharmaceutical chemistry;
C = Pharmaceutics; *D* = Pharmacognosy;
E = Pharmacy practice; *F* = Dispensing.

example, the subject average could be assigned to *F*, and the class position to *e*, so that the score of the first candidate gives a value of *a* of

$$a = \frac{30 - 20}{47.2} = 0.212$$

For the second candidate, $a = 0.763$, and so on, this is only one of an infinite number of possible sets of constants. Thus *F* could be given the arbitrary value of unity, and *e* could be the score in an intelligence test, carried out separately from the examination. These would give a different set of independent variables, but they would still give the scores shown in column *A* in Table 6.28.

The concept can be extended to the full diet of subjects by using equation (6.33) for the first candidate, and similar equations for the remainder.

$$X_1 = a_{1A}F_A + a_{1B}F_B + a_{1C}F_C + a_{1D}F_D + a_{1E}F_E + a_{1F}F_F \qquad (6.33)$$

These equations, embracing 20 dependent variables and 146 independent variables, are difficult to handle statistically, but the procedure can be simplified by

(a) Standardizing the scores in each column; these are given in Table 6.29. The means of the squares of the elements in each row are then equal to unity.

Table 6.29 — Standardized examination scores for pharmacy degree candidates

Candidate	Subject					
	A	*B*	*C*	*D*	*E*	*F*
1	−2.194	−0.581	−1.637	−1.460	−2.497	−1.050
2	0.102	−1.236	0.164	−0.312	0.353	0.137
3	0.612	−0.336	1.329	1.001	0.477	0.228
4	−0.791	−0.990	−0.365	0.755	0.105	−2.237
5	0.612	0.319	0.588	−0.804	0.725	0.776
6	1.122	0.728	1.329	1.412	−0.514	0.046
7	0.485	−1.236	−0.154	−1.132	0.849	0.776
8	−0.791	−1.481	−0.895	−0.230	−0.143	0.776
9	0.102	0.810	−0.048	0.180	0.105	0.594
10	1.122	2.365	1.118	1.412	1.221	1.598
11	−0.408	0.401	−0.260	−1.050	−0.390	−0.228
12	0.485	0.892	0.270	1.247	0.601	−0.411
13	1.378	0.892	1.647	0.591	1.592	1.689
14	0.102	−0.172	1.531	−0.722	−0.514	−0.868
15	1.250	1.056	1.118	1.247	1.221	1.324
16	0.740	−0.008	−0.471	0.016	−0.390	−0.776
17	−0.663	−0.827	−0.577	−1.132	−0.143	−1.050
18	−1.557	−1.236	−1.001	−1.378	−0.638	−0.685
19	−1.429	0.074	−1.319	−0.476	−2.001	−0.594
20	−0.281	0.565	0.694	0.837	−0.019	−0.046

A = Pharmacology; *B* = Pharmaceutical chemistry;
C = Pharmaceutics; *D* = Pharmacognosy;
E = Pharmacy practice; *F* = Dispensing.

(b) Standardizing the *F* values along the rows.
(c) Considering the results in terms of the covariance matrix of the standardized results.

Table 6.30 is the covariance matrix of the standardized version of the data, shown in Table 6.29. The matrix illustrates the original observation which stimulated Spearman's idea, thus if the leading diagonal results are ignored, the ratios of the numbers in any pair of rows are approximately constant. For rows A and C, for example, columns A and C are ignored, to give the remaining ratios.

$$\frac{0.5819}{0.5789}=1.0\ ; \qquad \frac{0.6424}{0.7398}=0.9\ ; \qquad \frac{0.9038}{0.7576}=1.2\ ; \qquad \frac{0.6426}{0.6454}=1.0$$

If 20 examination papers had been attempted by each of the 20 candidates, there would be 20 simultaneous equations analogous to equation (6.33), but each having 20, aF terms on the right-hand side. This would yield a square matrix containing $20 \times 20 = 400$ values of aF, each equal to the score of a particular candidate on a particular paper. If these values of aF are completely independent, principal components analysis will yield 400 eigenvectors and 20 eigenvalues, which will explain all the variation. The e term will therefore disappear. If some of the columns in the square matrix are related, the corresponding number of eigenvalues will reduce to zero, and less than 20 principal components will be necessary to completely describe the data. Factor analysis is necessary when the number of papers attempted is less than the number of candidates, and a square matrix does not result. A term e must now be introduced for each candidate, to make up the shortfall.

Summarizing therefore, factor analysis can be described as a version of principal components analysis in which the number of factors is less than the number of subjects assessing those factors, and which aims to reduce the number of factors to as few as possible.

Considering the (20×20) matrix described above, if the scores for the first candidate are called $x_{1,A}$ to $x_{1,T}$, and the scores for subject A are called $x_{1,A}$ to $x_{20,A}$, and so on, principal components analysis will give the matrix.

$$\begin{matrix} b_{1,A}x_{1,A} & b_{1,B}x_{1,B} & \ldots & b_{1,T}x_{1,T} \\ b_{2,A}x_{2,A} & b_{2,B}x_{2,B} & \ldots & b_{2,T}x_{2,T} \\ b_{20,A}x_{20,A} & \ldots & b_{20,T}x_{20,T} & \end{matrix}$$

in which the bs are the eigenvectors. Thus, reading along the rows, there will be 20 principal components (Z), as shown in equations (6.34) to (6.53)

$$Z_1 = b_{1,A}x_{1,A} + b_{1,B}x_{1,B} + \ldots + b_{1,T}x_{1,T} \tag{6.34}$$

$$Z_2 = b_{2,A}x_{2,A} + b_{2,B}x_{2,B} + \ldots + b_{2,T}x_{2,T} \tag{6.35}$$

$$\vdots \qquad\qquad\qquad \vdots$$

to

$$Z_{20} = b_{20,A}x_{20,A} + b_{20,B}x_{20,B} + \ldots + b_{20,T}x_{20,T} \tag{6.53}$$

Table 6.30 — Covariance matrix of standardized examination scores for pharmacy degree candidates

	A	*B*	*C*	*D*	*E*	*F*
A	1.0000	0.5819	0.8142	0.6424	0.9038	0.6426
B	0.5819	1.0000	0.5789	0.6256	0.3505	0.4874
C	0.8142	0.5789	1.0000	0.7398	0.7576	0.6454
D	0.6424	0.6256	0.7398	1.0000	0.5032	0.3234
E	0.8038	0.3509	0.7576	0.5032	1.0000	0.6379
F	0.6426	0.4874	0.6454	0.3234	0.6479	1.0000

A = Pharmacology; B = Pharmaceutical chemistry;
C = Pharmaceutics; D = Pharmacognosy;
E = Pharmacy practice; F = Dispensing.

Similarly, reading down the columns, there will be 20 equations in x, represented by equations (6.54) to (6.73)

$$x_A = b_{1,A}Z_{1,A} + b_{2,A}Z_{2,A} + \ldots + b_{20,A}Z_{20,A} \tag{6.54}$$

$$x_B = b_{1,B}Z_{1,B} + b_{2,B}Z_{2,B} + \ldots + b_{20,B}Z_{20,B} \tag{6.55}$$

$$\vdots \qquad\qquad \vdots$$

$$x_T = b_{1,T}Z_{1,T} + b_{2,T}Z_{2,T} + \ldots + b_{20,T}Z_{20,T} \tag{6.73}$$

For a factor analysis, there are less than the complete number of factors, 20 in this case. Terms are shed from the end of each equation, and substituted with an overall factor, e It is known that the sums of the squares of the eigenvectors in a row or a column are equal to unity, so that the value of e can be calculated from equation (6.74)

$$e = 1 - S(b^2) \tag{6.74}$$

Transportation of the equations to the form of equation (6.33) can be obtained by scaling the principal components so as to have unit variances. This is done by dividing the values of Z by their standard deviations, which are equal to the square roots of the corresponding eigenvalues, i.e.

$$F = Z/\sqrt{\lambda} \tag{6.75}$$

For a factor analysis of the information in Table 6.29, it will only be possible to obtain six principal components, because only 6 subjects have been considered. An estimate is possible by using eigenvectors calculated from the correlation matrix shown in Table 6.30. These are given in Table 6.31, together with the eigenvalues.

Table 6.32 — Eigenvectors and eigenvalues of standardized examination scores for pharmacy degree candidates

Eigenvalues	Eigenvectors					
4.082	−0.454	−0.356	−0.459	−0.385	−0.413	−0.374
0.816	−0.112	0.543	0.000	0.543	−0.440	−0.452
0.564	0.135	−0.589	0.195	0.384	0.320	−0.589
0.249	−0.518	−0.368	0.357	0.401	−0.304	0.464
0.150	0.547	−0.184	0.442	−0.265	−0.631	−0.061
0.138	−0.444	0.251	0.655	−0.423	0.207	−0.299

Substitution of the eigenvectors for b, and $F\sqrt{\lambda}$ for Z then gives equations (6.76) to (6.81), for example, the first term on the right-hand side of equation (6.76) is equal to $-0.454 \times \sqrt{4.082} = -0.92$. Only the first three factors are considered, because the eigenvalues indicate that these alone will explain over 90% of the variation, i.e.

$$\frac{(4.082+0.816+0.564)\times 100}{4.082+0.816+0.564+0.249+0.150+0.138} = 91.0\%$$

The eigenvalues in Table 6.31 indicate that 68% of the variation is explained by one principal component, and 91% by the first three. In these three equations (6.76) to (6.78), apart from the first term on the right-hand side, only one factor loading (0.49) is near 0.5, and none exceeds it.

$$X_1 = -0.92F_A - 0.10F_B + 0.10F_C + 0.13 \quad (6.76)$$

$$X_2 = -0.72F_A + 0.49F_B - 0.44F_C + 0.05 \quad (6.77)$$

$$X_3 = -0.93F_A - 0.00F_B + 0.15F_C + 0.11 \quad (6.78)$$

$$X_4 = -0.77F_A + 0.49F_B + 0.29F_C + 0.08 \quad (6.79)$$

$$X_5 = -0.83F_A - 0.40F_B + 0.24F_C + 0.09 \quad (6.80)$$

$$X_6 = -0.76F_A - 0.41F_B - 0.44F_C + 0.06 \quad (6.81)$$

The conclusion therefore is that the overall abilities of most of the candidates can be estimated from their performances in one subject. However, rotation of the first two principal components, a process which will be explained later (page 125), gives a clearer picture of the situation, as shown in equations (6.82) and (6.83).

The asterisks indicate rotated values.

$$x_1{}^* = -1.05F_A{}^* + 0.01F_B{}^* + 0.01F_C{}^* \quad (6.82)$$

$$x_2{}^* = -0.51F_A{}^* + 0.50F_B{}^* - 0.45F_C{}^* \qquad (6.83)$$

Thus, the first principal component depends on only one subject result, and assesses the overall performances of 4.082 × 100/6, nearly 70% of the candidates, while the seond principal component is a composite function to which the performances in three subjects have roughly equal contributions, and explains a further 14%. In contrast, the last two principal components are responsible for only 5%. It therefore follows that the number of principal components taken into consideration depends on the probability level which is acceptable.

ROTATION

The loadings obtained in a factor analysis exercise form one of an infinite number of possible sets of values, all of which fit the data. If we consider the point P in Fig. 6.12,

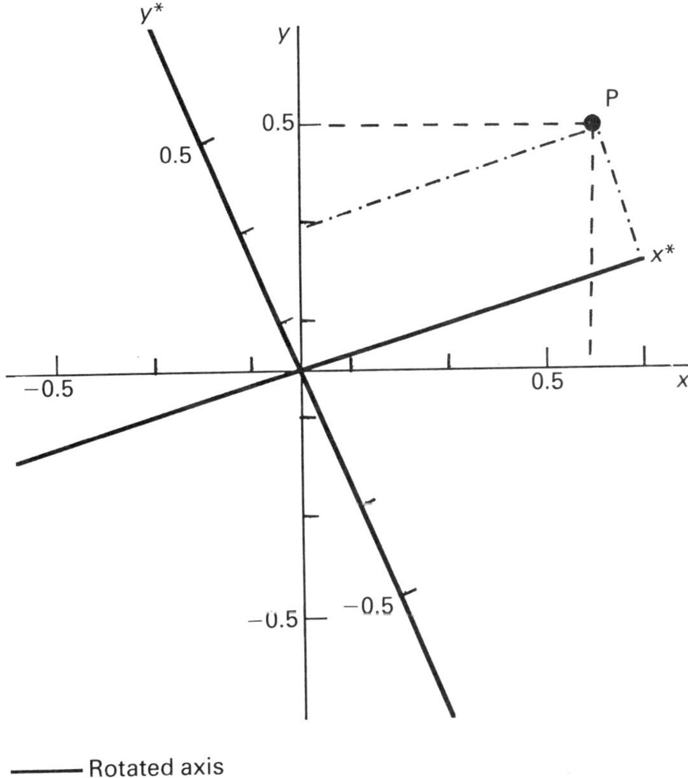

Fig. 6.12 — Rotation procedure.

its coordinates are determined by the lengths of the perpendiculars from the axes to the point, and are equal to (0.6, 0.5). If the origin is left in its original position, but the axes are rotated around the origin, the coordinates will change; for example, if the axes are rotated through 25°, as shown in Fig. 6.12, the new coordinates will be (0.76,

0.19). However, the positions of the origin and the point P are unchanged, so that the direction and length of the vector OP remain the same. It can be shown that for an anticlockwise rotation through the angle ϕ, the original coordinate (x) is related to the rotated coordinate (x^*) through equation (6.84), and the corresponding coordinates for y are related through equation (6.85).

$$x^* = x\cos\phi + y\sin\phi \tag{6.84}$$

$$y^* = -x\sin\phi + y\cos\phi \tag{6.85}$$

The equations can be expressed in matrix algebra as follows,

$$\begin{bmatrix} x^* \\ y^* \end{bmatrix} = \begin{bmatrix} \cos\phi & \sin\phi \\ -\sin\phi & \cos\phi \end{bmatrix} \begin{bmatrix} x \\ y \end{bmatrix} \tag{6.86}$$

A similar operation can be carried out in three dimensions for equations having the form,

$$z = ax + by \tag{6.87}$$

in which a and b are constants, and x, y and z are variables.

Once again the origin does not move, but the axes are rotated through the vertical and horizontal planes, and the process can be demonstrated visually using a three-dimensional model. The number of dimensions can be extended beyond three, but the resulting systems cannot be demonstrated in visual space.

The object of rotation is to change coordinates so that they either approach zero, or plus or minus one, and to eliminate intermediate values. The simplest situation is when, in a set of principal components, all but one of the terms in each of the components is reduced to zero. The exception should have a value near to plus or minus unity, and should involve a different factor for each principal component. Thus for example, if in equations (6.88) and (6.89) the factor loadings are around 0.7 ($0.7^2 = 0.5$), all that can be deduced is that both X_1 and X_2 are influenced in some way by F_A *and* F_B, but a rotation which reduces a_{1A} *and* a_{2B} to about zero, and increases a_{1B} and a_{2A} to numbers around ± 1.0 will transform the equations to equations (6.90) and (6.91)

$$X_1 = a_{1A}F_A + a_{1B}F_B \tag{6.88}$$

$$X_2 = a_{2A}F_A + a_{2B}F_B \tag{6.89}$$

$$X_1{}^* = a_{1B}F_B \tag{6.90}$$

$$X_2{}^* = a_{2A}F_A \tag{6.91}$$

These rotated equations indicate that $X_1{}^*$ is dependent on F_B but is independent of F_A, and vice versa, thereby simplifying the interpretation process.

The examination results of pharmacy candidates given above (page 120) are an example of the usefulness of rotation. 81% of the results can be explained by equation (6.76), which suggests that X_1 is dominated by the term F_A, while X_2 is also mainly dependent on F_A, but F_B and F_C could be making significant contributions. A plot of the coefficients of X_1 against X_2 gives three points, A (-0.92, -0.72), B

$(-0.10, 0.49)$ and $C\,(0.10,\ -0.44)$, shown in Fig. 6.13, and representing the F_A, F_B and F_C terms respectively. The vector **B** lies at an angle of $\tan^{-1}(0.10/0.49) = 11.5°$ and **C** at an angle of $\tan^{-1}(0.10)/(0.44) = 12.8°$, giving a mean angle of 12.2°. Rotation of the axes through this angle, as shown in Fig. 6.13, gives new coordinates, as follows,

$$\mathbf{A}\begin{bmatrix} X_1{}^* \\ X_2{}^* \end{bmatrix} = \begin{bmatrix} \cos 12.2° & \sin 12.2° \\ -\sin 12.2° & \cos 12.2° \end{bmatrix}\begin{bmatrix} X_1 \\ X_2 \end{bmatrix} = \begin{bmatrix} 0.977 & 0.211 \\ 0.211 & 0.977 \end{bmatrix}\begin{bmatrix} -0.92 \\ -0.72 \end{bmatrix}$$

$$= \begin{matrix} -0.92 \times 0.977 + -0.72 \times 0.211 \\ -0.92 \times 0.211 + -0.72 \times 0.977 \end{matrix} = \begin{bmatrix} -1.05 \\ -0.51 \end{bmatrix}$$

$$\mathbf{B}\begin{bmatrix} X_1{}^* \\ X_2{}^* \end{bmatrix} = \begin{bmatrix} 0.977 & 0.211 \\ -0.211 & 0.977 \end{bmatrix}\begin{bmatrix} -0.10 \\ 0.49 \end{bmatrix} = \begin{bmatrix} 0.006 \\ 0.50 \end{bmatrix}$$

$$\mathbf{C}\begin{bmatrix} X_1{}^* \\ X_2{}^* \end{bmatrix} = \begin{bmatrix} 0.977 & 0.211 \\ -0.211 & 0.977 \end{bmatrix}\begin{bmatrix} 0.10 \\ -0.44 \end{bmatrix} = \begin{bmatrix} 0.005 \\ -0.45 \end{bmatrix}$$

These give the rotated equations (6.82) and (6.83), which clearly show that $X_1{}^*$ is completely dependent on F_A, but $X_2{}^*$ is influenced approximately equally by F_A, F_B and F_C. This is confirmed by Fig. 6.13, which shows that both B and C lie on the $X_2{}^*$ axis, contributing nothing to the value of $X_1{}^*$, but having large $X_2{}^*$ coordinates, thereby contributing to the value of $X_2{}^*$.

Baines (1978) carried out a factor analysis on 15 toothpaste flavours, using a panel of 12 'semi experts', who awarded scores on the 11 attributes listed in Table 6.32, together with the overall flavour preference. His treatment of the data, slightly modified, is described below. Principal component analysis yielded a set of eigenvalues on which the first three accounted for the bulk of the variance of the results. The remainder were therefore rejected. The eigenvalues of these, designated I, II and III, are given in Table 6.32. Relationship patterns can be recognized by plotting these eigenvectors against each other, for example, flavour strength has the coordinates 0.83, 0.14 and -0.46 in three-dimensional space. A solid model of this sort is difficult to handle, and was simplified by plotting II/I against III/I, as shown in Fig. 6.14. The points roughly form a triangle, the corners of which are occupied by the attributes 4, 8 and 10, 1, 3 and 6, and 6 and 11. It was therefore suggested that the corners of the triangle represented 'pure sensations', while the sides of the triangle are mixtures of pure sensations. Trial and error suggests that 3, 4 and 11 are the best representatives of pure sensations. Calculation of the correlation coefficient between sweetness and warming ($r = 0.019$) indicates that the two attributes are not related, adding credance to their choice as pure sensations. Lasting freshness also has low correlation coefficients with both sweetness and warming ($r = 0.139$ and 0.293), making it a good candidate for the third corner.

A set of eigenvectors, as exemplified by columns I to III in Table 6.32, is one of an infinite choice of combinations of numbers which can represent the data. If, for

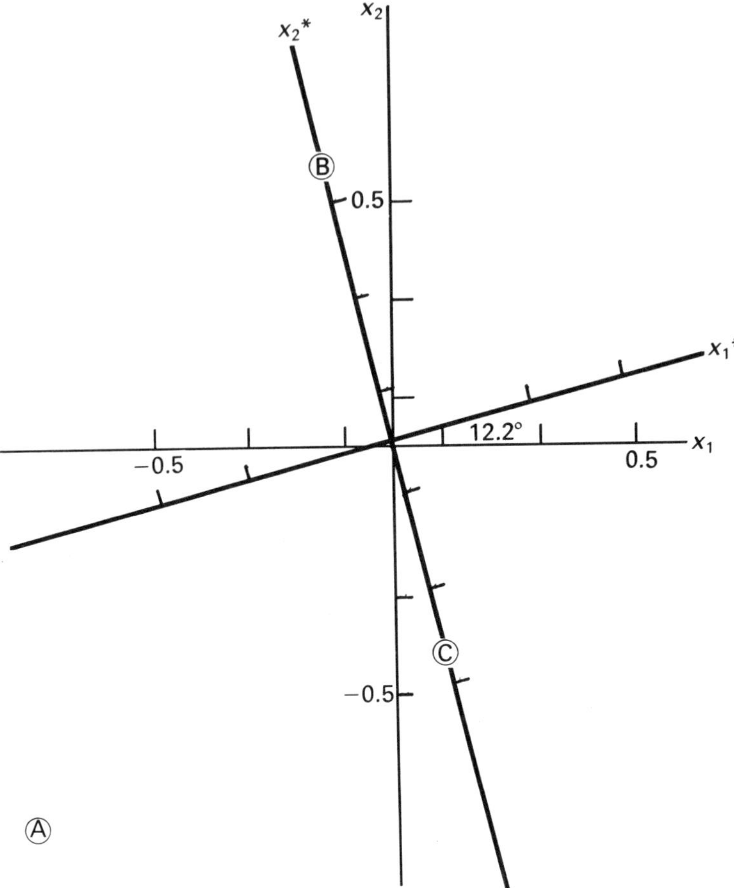

Fig. 6.13 — Rotation of axes for pharmacy degree examination scores.

example, I is plotted against III, as shown in Fig. 6.15, the coordinates of the points can be varied continuously by rotating the axes through 360°. Two of the points representing the corners of the triangle in Fig. 6.14, namely 3 and 4, are so placed that they can be made to coincide with the axes, if the axes are rotated through the necessary angle. This can be done by multiplying the coordinates by

$$\begin{bmatrix} \cos\phi & \sin\phi \\ -\sin\phi & \cos\phi \end{bmatrix}$$

where ϕ is the angle through which the axes are rotated. The background to this process is explained elsewhere (page 125). The process must be implemented so that the 'pure' attributes, warming, sweetness and lasting freshness each lie on or near to

Table 6.32 — Principal components analysis of toothpaste flavour results

Attribute	Principal components							
	Unrotated				Rotated			
	I	II	III	Commu-nality	I*	II*	III*	Cummu-nality
1. Flavour stength	0.83	0.14	−0.46	0.92	0.319	0.026	0.909	0.93
2. Tingling	0.88	−0.15	−0.29	0.88	0.357	−0.299	0.819	0.89
3. Warming	0.57	0.13	−0.54	0.63	0.088	0.105	0.788	0.64
4. Sweetness	−0.61	−0.13	−0.65	0.81	−0.878	0.201	0.105	0.82
5. Mildness	−0.88	−0.31	0.16	0.90	0.606	−0.097	0.704	0.90
6. Freshness	0.62	−0.67	0.21	0.88	0.315	−0.838	−0.275	0.88
7. Sharpness	0.90	0.14	−0.16	0.86	0.559	−0.067	0.738	0.86
8. Bitterness	0.74	0.23	0.52	0.87	0.919	−0.110	0.131	0.87
9. Lasting flavour	0.44	−0.45	−0.36	0.53	−0.094	−0.445	−0.566	0.53
10. Lasting bitterness	0.69	0.34	0.53	0.87	0.931	0.003	0.089	0.87
11. Lasting freshness	0.47	−0.77	0.26	0.88	0.234	−0.904	−0.135	0.89

one of the axes, in three-dimensional space, and new principal components calculated so that they lie along these axes. The plot of I against III is shown in Fig. 6.15. For the point '3' (0.57, −0.54) to coincide with the III axis, it will be necessary to rotate the axes through an angle of $\tan^{-1}(0.54/0.57) = 43.5°$. Similarly, to bring the point '4' on to the I axis, the axes must be rotated through $\tan^{-1}(0.61/0.65) = 43.2°$.

The mean is 43.3°. To rotate each point clockwise through this angle, the coordinates must be multiplied by the matrix,

$$\begin{bmatrix} \cos 43.3° & \sin 43.3° \\ -\sin 43.3° & \cos 43.3° \end{bmatrix} \quad \text{or} \quad \begin{bmatrix} 0.62 & 0.78 \\ -0.78 & 0.62 \end{bmatrix}$$

Thus for point '3' (0.57, −0.54), the new coordinates (I*, III*) are given by

$$\begin{bmatrix} \text{I}^* \\ \text{III}^* \end{bmatrix} = \begin{bmatrix} 0.62 & 0.78 \\ -0.78 & 0.62 \end{bmatrix} \begin{bmatrix} 0.57 \\ -0.54 \end{bmatrix} = \begin{bmatrix} -0.07 \\ -0.78 \end{bmatrix}$$

The rotated plot is shown in Fig. 6.15. It will be noted that the points '3' and '4' are close to the ordinate and abscissa respectively.

It is now necessary to rotate point 11 on to one of the other two planes in three-dimensional space. This is done by plotting either I* or III* against II, and rotating the axes so that point 11 lies on the II axis.

Rotation of I against II is shown in Fig. 6.16, and suggests a rotation angle of

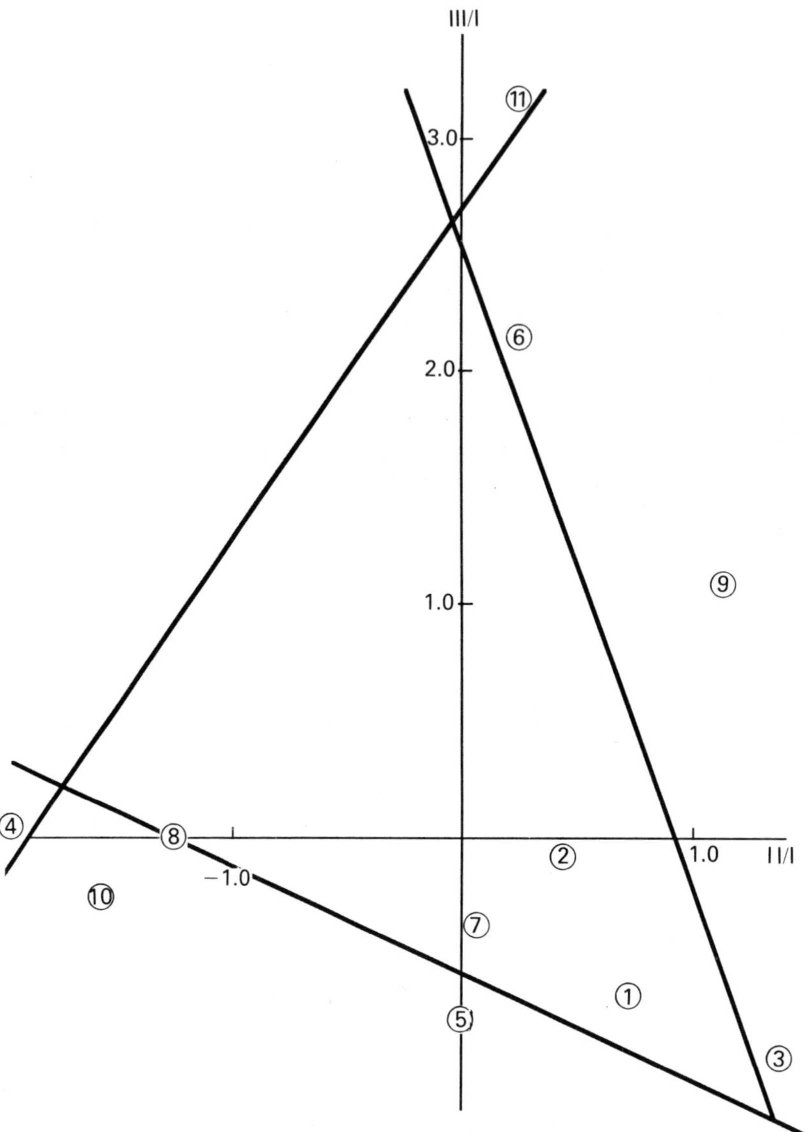

Fig. 6.14 — Evaluation of toothpaste flavours.

16.9°, which brings point 11 on to the coordinates (0.00, −0.08), and points '3' and '4' on to (0.12, 0.10) and (−0.88, 0.13) respectively. The rotated values of I*, II* and III* are given in Table 6.32. Communalities are virtually unchanged from the unrotated values.

If the I* and II* axes are considered to form the vertical plane, the I* axis in this plane forms a scale varying from sweetness (4) and mildness (5), having negative

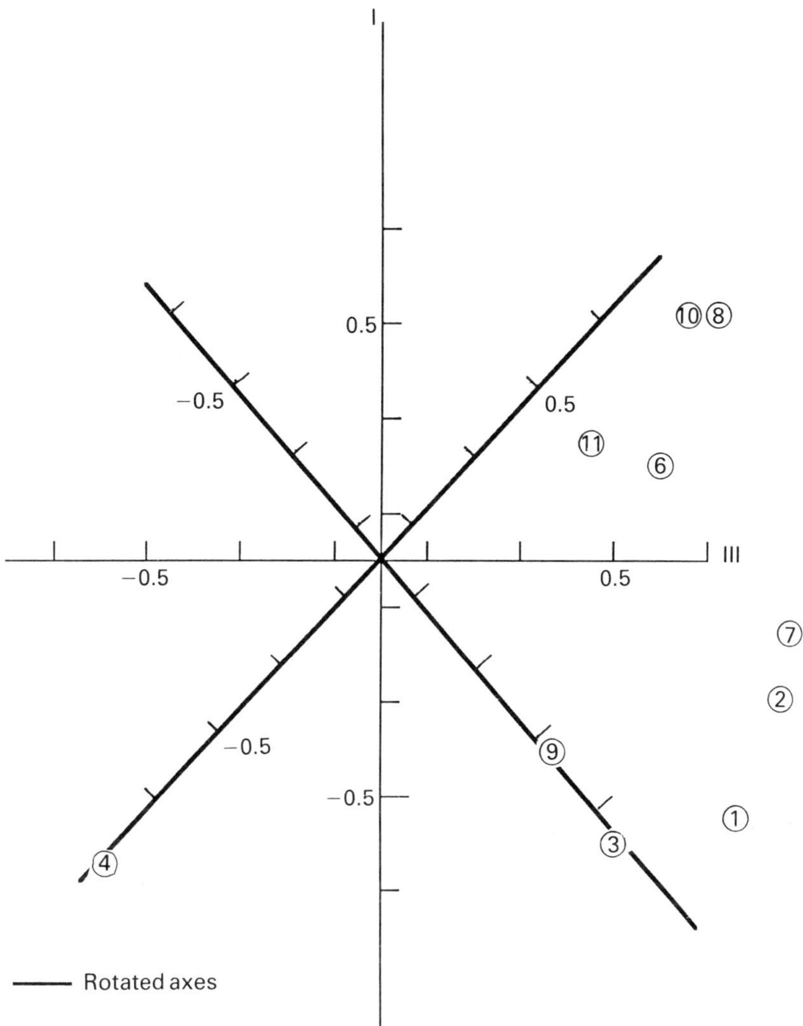

Fig. 6.15 — I and II components for toothpaste evaluation.

values, to bitterness (8) on the extreme positive side. Lasting bitterness (10), not surprisingly, lies close to bitterness (8). Flavour (1), warming (3), freshness (6) and sharpness (7) also have positive values. All lie close to the I* axis, so that their contributions to II* are small. Lasting flavour (9) and lasting freshness (11) lie on the II* axis, which must therefore be assumed to be a measure of duration of sensation. Warming (3) gives a significant contribution to only III*, which it shares with lasting flavour (9). One could possibly classify these factors as 'feeling sensations', although the place of (9) in this factor is obscure. Tingling (2) makes a contribution to II*, but it also contributes significantly to the other two factors.

Thus, 11 attributes have been resolved into three factors, most of the attributes giving their major thrust to only one factor.

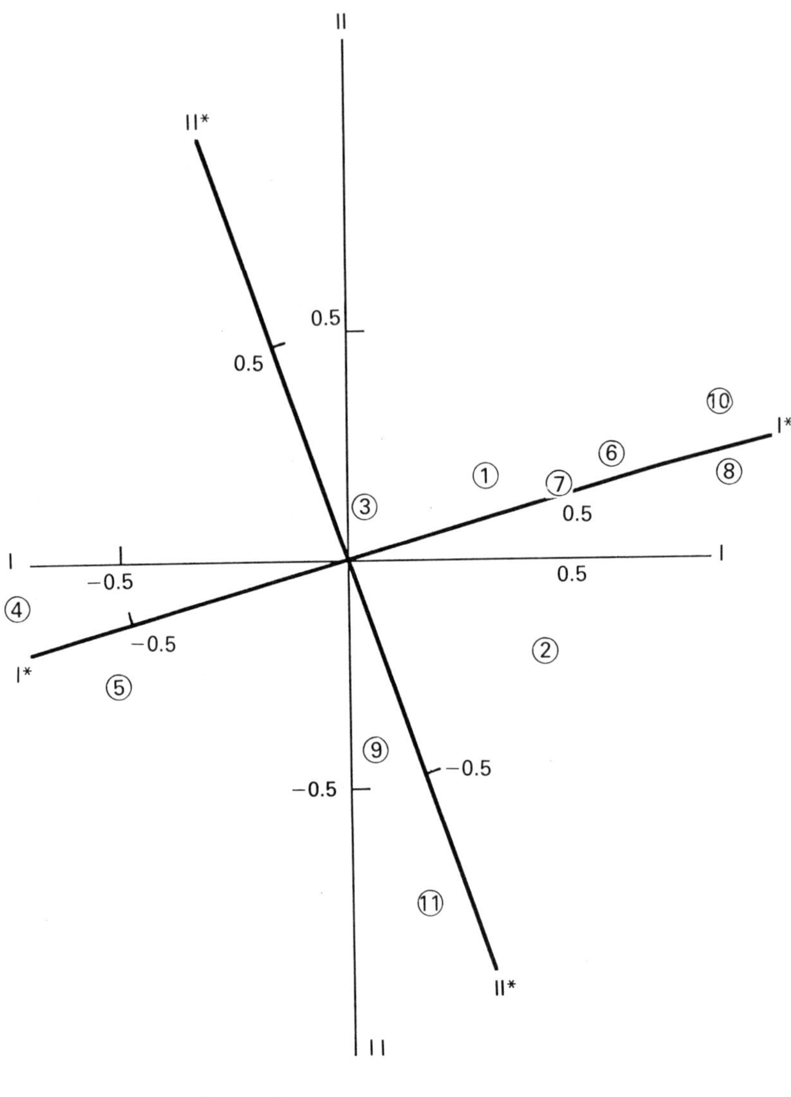

Fig. 6.16 — I* and II* coordinates of toothpaste evaluation.

COMPUTERIZED ROTATION

A procedure of the type described above would be too protracted for complicated systems, so computerized methods have been devised for establishing the rotation required to facilitate easy interpretation. A typical procedure can be explained by considering a vector OP in the plane of two factors F_A and F_B, as shown in Fig. 6.17, forming a small angle (ϕ) with the F_A axis. Rotation of the axes is equivalent to moving the point P along a circular path, with the origin at its centre, so that the line

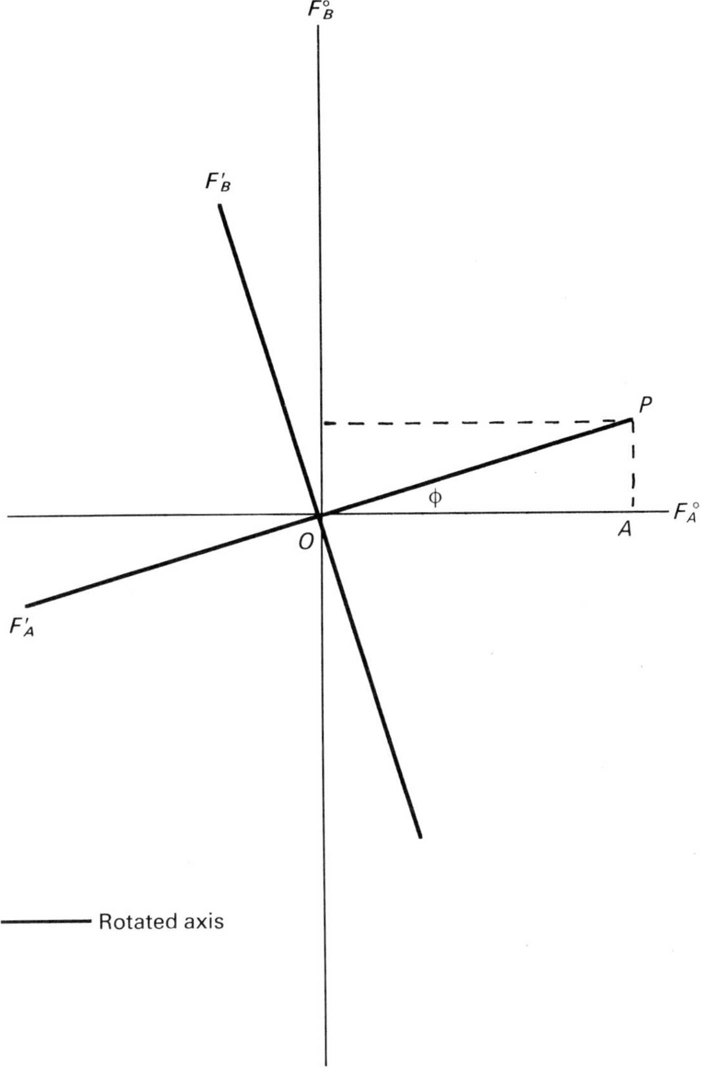

Fig. 6.17 — Computerized treatment of rotation.

AP will be an arc of the circle. However, since ϕ is small, the arc can be approximated to a straight line, normal to the F_A axis, forming a right angled triangle AOP with the vector and the F_A axis. If the axes are rotated through the angle ϕ, the F_A axis will coincide with the vector, and the rotated coordinates of the point P will be $F'_A = h$, the length of the vector, and $F'_B = 0$. The original coordinates are given by equation (6.92) and (6.93).

$$\mathrm{OA} = \mathrm{h}\cos\phi = F^\circ{}_A \qquad (6.92)$$

$$AP = h\sin\phi = F^{\circ}_{B} \tag{6.93}$$

so that

$$F^{\circ}_{A}F^{\circ}_{B} = h^2 \sin\phi\cos\phi = 0.5h^2\sin 2\phi \tag{6.94}$$

also,

$$F^{\circ 2}_{A} + F^{\circ 2}_{B} = h^2 \tag{6.95}$$

therefore,

$$h^4 = F^{\circ 4}_{A} + F^{\circ 4}_{B} + 2F^{\circ}_{A}F^{\circ}_{B} \tag{6.96}$$

$$= F^{\circ 4}_{A} + F^{\circ 4}_{B} + h^4\sin^2 2\phi \tag{6.97}$$

Since h (and therefore h^4) is constant, as ϕ (and therefore $\sin^2 2\phi$) decreases, the value of $F^{\circ 4}_{A} + F^{\circ 4}_{B}$ must increase, and when one of the axes coincides with the vector, $\phi = \sin^2 2\phi = 0$ and the value of $F^{\circ 4}_{A} + F^{\circ 4}_{B}$ will be maximal. The varimax (or quartimax) procedure, which is said to maximize the results, assesses the rotation angle which gives the maximum sum of the fourth powers of the data matrix.

REFERENCES

Armstrong, N. A., Gebre-Mariam, T., James, K. C. and Kearney, P. (1987) The influence of viscosity on the migration of chloramphenicol and 4-hydroxybenzoic acid through glycerogelatin gels *J. Pharm. Pharmacol.* **39** 583–586.

Baines, E. (1978) Factor analysis in the evaluation of cosmetic products. *Proceedings of a Symposium on Product Evaluation*, Society of Cosmetic Scientists, Eastbourne.

Bate-Smith, E. C. and Westall, R. G. (1950) Chromatographic behaviour and chemical structure. 1. Some naturally occurring phenolic substances. *Biochem. Biophys. Acta* **4** 427–440.

Evans, R. M. and Farr, S. J. (Unpublished results).

Hammett, L. P. (1940) Physical Organic Chemistry, McGraw-Hill, New York and London, p. 186.

Hansch, C., Unger, S. H. and Forsythe, A. B. (1973) Strategy in drug design. Cluster analysis as an aid in the selection of substituents. *J. Med. Chem.* **16** 1217–1222.

Hansch, C. and Leo, A. J. (1979) *Substituent Constants for Correlation Analysis* in *Chemistry and Biology*, John Wiley, New York.

Harris, A. J., James, K. C., Powell, M. and Bishop, G. B. (1975) Assessment of auxilliary detergents in shampoo mixtures. *Cosm. Perf.* **90**(10) 23–25, 28, 30.

Iwasa, J., Fujita, T. and Hansch, C. (1965) *J. Med. Chem.* **8** 150–153 (Reviews on Hansch substituent constants can be found in James 1974, 1988 and Hansch and Leo 1979).

James, K. C. (1988) In Introduction to the *Principles of Drug Design* (2 edn.), (ed. Smith, H. J.) Wright, London, pp. 240–264.

James, K. C. (1974) In *Progress in Medicinal Chemistry*, Vol. 10 (eds. Ellis, G. P. and West, G. B.) Elsevier, pp. 205–243.

James, K. C., Nicholls, P. J. and Richards, G. T. (1975) Correlation of androgenic activities of the lower testosterone esters with R_m values and hydrolysis rates. *Eur. J. Med. Chem.* **10** 55–58.

McFarland, J. W. and Gans, D. J. (1986) The significance of clusters in the graphical display of structure-activity relationships. *J. Med. Chem.* **29** 505–514.

Powell, M., James, K. C., Harris, A. J. and Bishop, G. B. (1973) An assessment of auxiliary detergents in shampoo mixtures. *Proceedings of a Symposium on Evaluation of Product Performance*, Society of Cosmetic Chemists of Great Britain, Nottingham.

Rushton, D. H. (1988) Chemical and morphological properties of scalp hair. Ph.D.Thesis, University of Wales.

Spearman, C. (1904) General intelligence. Objectivity determined and measured. *Am. J. Psych.* **15** 201–293.

Taft, R. W. (1956) In Steric Effects in Organic Chemistry (ed. Newman, M. S.), Wiley, New York, p. 587.

Swain, C. G. and Lupton, E. C. (1968) Field and resonance components of substituent effects. *J. Am. Chem. Soc.* **90** 4328–4337.

ADDITIONAL READING

Coulson, A. E. (1965) *Introduction to Matrices*, Longman Group Ltd., London.

Manly, B. F. J. (1986) *Multivariate Statistical Methods*. A Primer, Chapman and Hall, London and New York.

7

Optimization

MODEL-DEPENDENT OPTIMIZATION

The design of pharmaceutical products and processes often involves a compromise between two or more conflicting factors. For example, tablets must be strong enough to withstand the rigours of packaging, handling and transport. Yet at the same time they must comply with pharmacopoeial standards for disintegration and dissolution, and both of these are adversely affected when the physical strength of the tablet is increased. Also, in virtually every process, time and/or cost are limiting factors. Thus an optimum is required, which is the best possible compromise in the given circumstances. In general terms, the optimum solution may be those values of two or more experimental variables which together give the 'best', i.e. the maximum or minimum possible value for a dependent variable.

The procedure can best be shown by using an example which will initially be simple in its design, and then will be rendered more complex to show the power of the technique. Suppose it is required to produce tablets which are as strong as possible. Yet the formulation requires a disintegrant, and it is known that disintegrants cause a reduction in tablet strength. Thus the problem is to find the combination of compression pressure and disintegrant which will give the strongest tablets possible yet still comply with standards for tablet disintegration.

In this example, compression pressure and disintegrant concentration are the independent variables. These are the experimental conditions which are under the control of the experimenter. Tablet strength and disintegration time are the dependent variables. Their magnitudes are governed by the values of the independent variables.

Optimization situations may be classified into two types, unconstrained and constrained. Consider the following quotation:

'We shall defend our island, whatever the cost may be... we shall never surrender' (Winston Churchill, 4th June, 1940).

This is an unconstrained situation, as the objective is to be achieved unconditionally.

Consider another historic quotation:

'The US will land a manned spacecraft on the moon before the decade is out' (President John Kennedy, May 1961).

Here there are constraints or conditions. It is to be the US which lands a manned spacecraft on the moon and there is a time limit. (However the usual constraint on our actions, availability of finance, is noticeably absent!)

Constrained optimization problems are much more common, and the tabletting example is one in which there are upper and lower limits on the values which can be adopted by both independent variables. The compression pressure cannot be less than zero and an upper limit is dictated by the maximum pressure which can be exerted by a given tablet press. As far as the disintegrant concentration is concerned, again the minimum cannot be less than zero and the maximum must be less than 100% otherwise the disintegrant would comprise the whole tablet.

Optimization methods are an example of a series of techniques called 'response surface methodology', and can be likened to climbing to the top of a hill. The summit of a hill can be located in two ways. One method is to prepare a contour map, joining together points of equal altitude. Alternatively one can start at a point on the hill's surface, and by proceding ever upwards the highest point can be achieved. The latter is an analogy of model-independent optimization, and is discussed in Chapter 8. The use of a contour map is analogous to model-dependent optimization.

If the independent and dependent variables are connected by straight-line relationships, these are examples of linear programming. Linear programming was devised some years ago and is widely used as a management tool, but has received surprisingly little attention in the pharmaceutical literature. This is a pity since many pharmaceutical problems are problems of optimization and many pharmaceutical relationships are linear or can be made linear using simple mathematical transformations such as taking logarithms.

The first stage of the process is to obtain experimental data, and this is best achieved by means of a factorial design. Thus two compression pressures (X_1) and two disintegrant concentrations (X_2) are chosen, the tablets prepared and the tablet strengths (Y_1) and disintegration times (Y_2) measured. Results are given in Table 7.1. The experimental notation is that described for two-level factorials in Chapter 4.

Table 7.1 — The dependence of tablet strength and disintegration time on compression pressure and disintegrant concentration

Experiment —	Compression pressure (MPa) (X_1)	Disintegrant concentration (%) (X_2)	Disintegration time (sec) (Y_1)	Tablet strength (kg) (Y_2)
(1)	100	2.5	500	6.1
x_1	300	2.5	1070	9.4
x_2	100	7.5	140	4.9
x_1x_2	300	7.5	640	8.2

The data are presented graphically (Fig. 7.1a and 7.1b) by plotting Y_1 and Y_2 against X_1. In both cases, graphs showing two virtually parallel lines are obtained. This indicates that there is no significant interaction between the two independent variables.

The data can also be combined into two three-dimensional graphs. On these, the independent variables form the horizontal axes and the vertical axis is the dependent variable. (Figs. 7.2a and 7.2b). The rectangle ABCD is known as the response surface and gives the tablet strength or the disintegration time for any combination of compression pressure and disintegrant concentration.

Though the response surfaces of the two graphs look rather similar in overall shape, it must be borne in mind that the required characteristics of the tablet are high strength (a maximum) and low disintegration time (a minimum). Obviously therefore these two requirements are in conflict with each other, and an optimum solution must be sought.

The disintegration results will be considered first.

The next stage in the optimization procedure is to carry out multiple regression analysis. This involves fitting the values of a dependent variable and the independent variables into a polynomial equation of the form

$$Y_1 = B_0 + B_1X_1 + B_2X_2 \tag{7.1}$$

In this case, the dependent variable Y_1 is the disintegration time of the tablet.

Multiple regression analysis is used to obtain the values of the coefficients B_0, B_1 and B_2. Thus an expression is obtained linking the two independent variables to one of the dependent variables. Substitution of the data from Table 7.1 into equation (7.1) yields:

$$Y_1 = 447 + 2.67X_1 - 79.0X_2 \tag{7.2}$$

Increasing the compression pressure will increase disintegration time, as also will decreasing the disintegrant concentration. Hence the signs of the coefficients of X_1 and X_2 are positive and negative respectively. It is useful to check the signs of the coefficients at this stage.

Equation (7.2) can be rearranged to give:

$$X_1 = \frac{Y_1 - 447 + 79.0X_2}{2.67} \tag{7.3}$$

This enables the question to be answered 'If tablets of a given disintegration time, and containing a given concentration of disintegrant are required, what compression pressure is needed?' Thus Y_1 and X_2 are specified, B_0, B_1 and B_2 are known and the only unknown is X_1. Thus combinations of X_1 and X_2 can be obtained which will give any specified value of Y_1. If these are plotted, a series of parallel lines is obtained (Fig. 7.3).

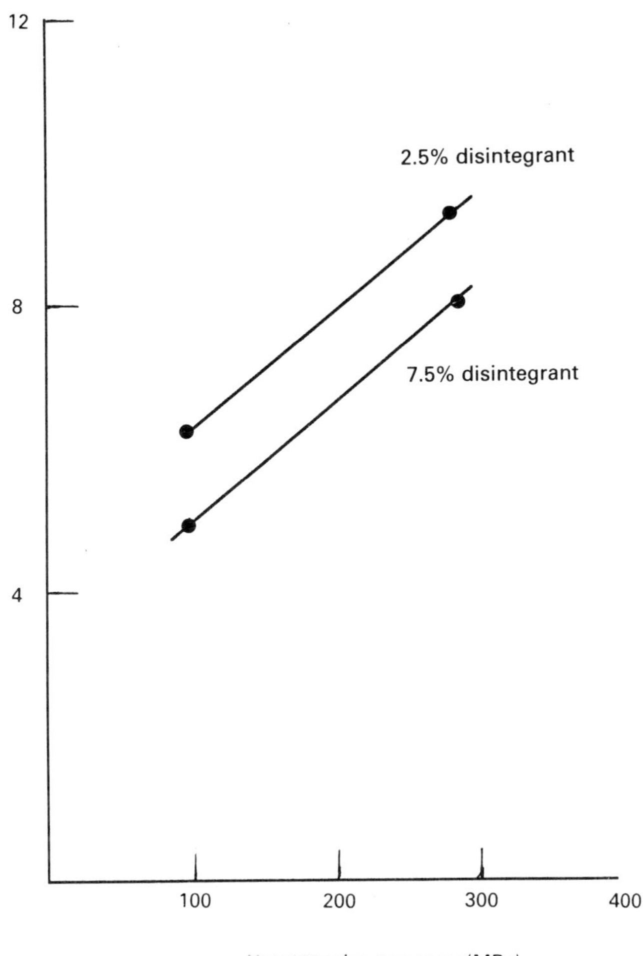

Fig. 7.1(a) — Relationship between compression pressure and tablet strength, using two concentrations of disintegrant.

With a linear relationship, there is no maximum value and there are hence an infinite number of combinations of the two independent variables which will give a specified value of the dependent variable. However, constraints can now be applied.

(1) X_1 and X_2 cannot be less than zero (these constraints represent the axes of Fig. 7.3).
(2) Y_1 cannot be greater than 900 seconds (BP limit for tablet disintegration time).
(3) X_1 cannot exceed the maximum pressure that the press can apply (say 400 MPa).
(4) X_2 cannot exceed a given concentration, limited by the formulation (say 10%).

Thus area ABCDE of Fig. 7.3 represents all combinations of compression

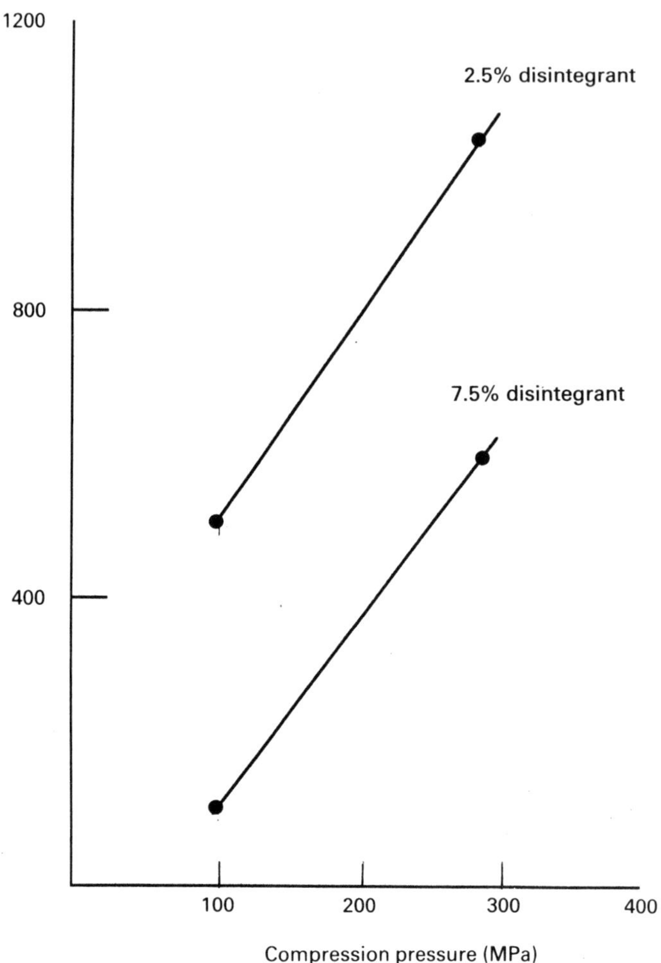

Fig. 7.1(b) — Relationship between compression pressure and tablet disintegration time, using two concentrations of disintegrant.

pressure and disintegrant concentration which will give tablets disintegrating in 900 seconds or less.

There is a great temptation to extrapolate these relationships, and to assume that the relationship holds outside the range of values of the independent variables which have been studied. In this case, a constraining boundary has been breached in the original design (a disintegration time of 900 seconds) and so the question of extrapolation does not arise here.

The same treatment can be used for the tablet strength data.

Multiple regression gives the equation

$$Y_2 = 5.05 + 0.0165X_1 - 0.240X_2 \qquad (7.4)$$

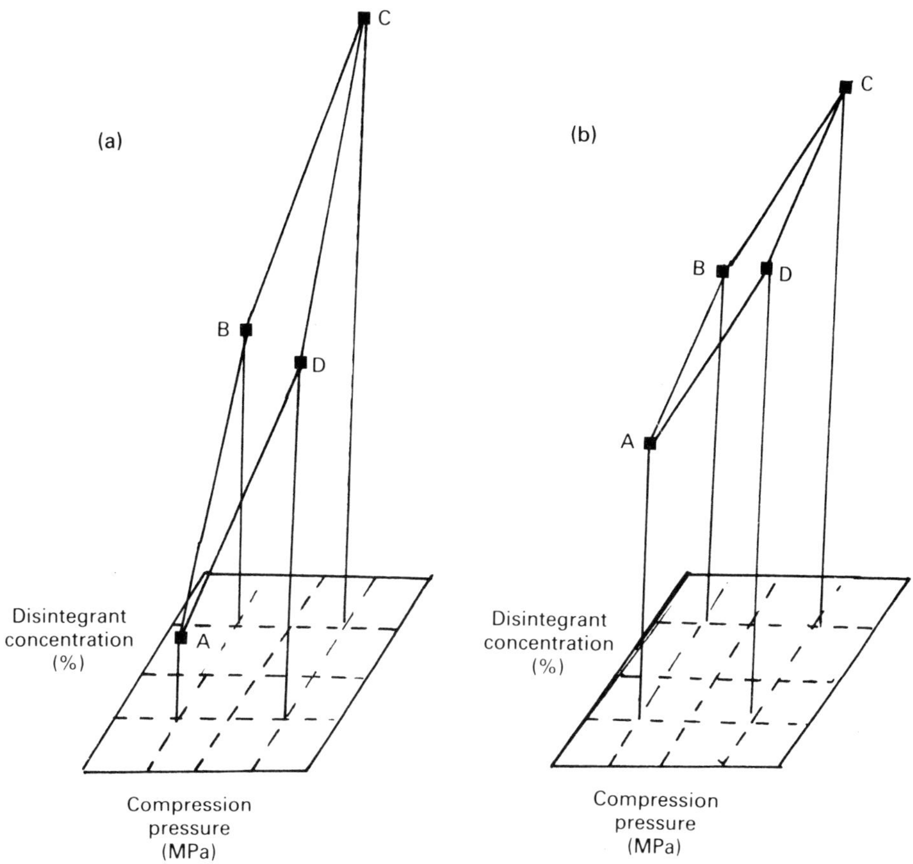

Fig. 7.2 — (a) Three-dimensional representation of the relationship between compression pressure, disintegrant concentration and tablet disintegration time. ABCD is the response surface. (b) Three-dimensional representation of the relationship between compression pressure, disintegrant concentration and tablet strength. ABCD is the response surface.

Rearrangement gives

$$X_1 = \frac{Y_2 - 5.05 + 0.240X_2}{0.0165} \quad (7.5)$$

Fig. 7.4 shows another series of parallel lines representing the values of compression pressure and disintegrant concentration which give a specified tablet strength.

The possibility must now be considered that the selected combinations of experimental conditions may not give tablets possessing the required properties. For example, imagine that there is a constraint to the effect that tablets must have a minimum tablet strength of 10 kg. This is greater than any of the tablets reported in Table 7.1. It is tempting to extrapolate, and calculate combinations of conditions

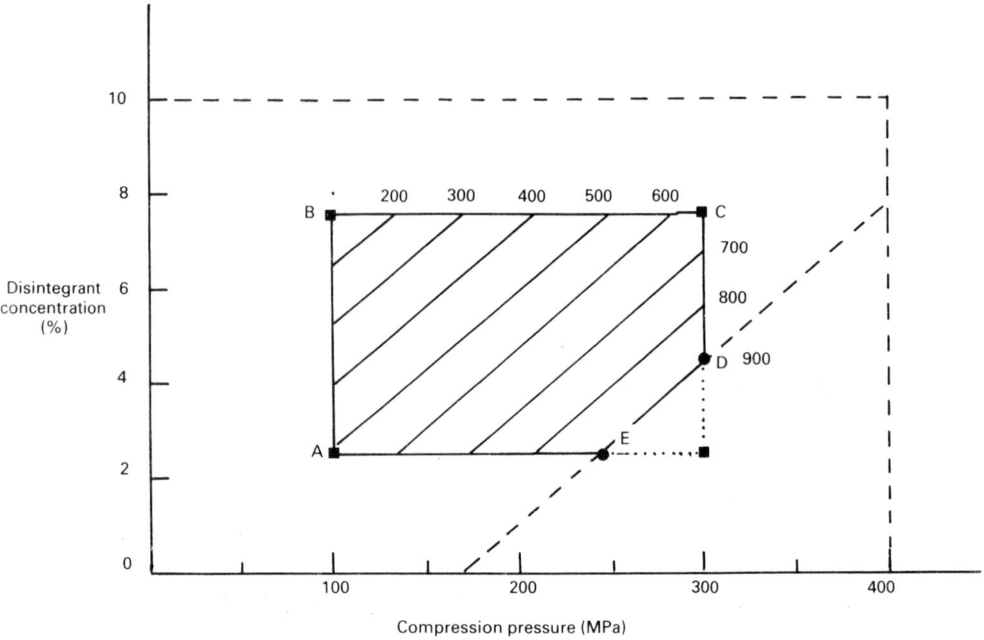

Fig. 7.3 — Contour plot of tablet disintegration time derived from a two-factor, two-level factorial design.

which will give tablets of the required strength. However, the inherent hazards of this approach must be borne in mind. A more satisfactory procedure is to extend the study. This is done by the use of one or more additional factorial designs.

Consider Fig. 7.5. A line is drawn passing through the centre point of the study and perpendicular to the series of parallel lines. This is the path of steepest ascent. The point of intersection between this line and the 'box' of the original design then forms the centre point of the next factorial design. This intersection is at point A, the coordinates of which are 300 MPa and 2.7%. Thus a suitable second factorial design would be to use pressures of 250 and 350 MPa and disintegrant concentrations of 1.7 and 3.7%, as shown in Fig. 7.5.

The preceding treatment deals with the two dependent variables separately, and as such gives the extreme values of the disintegration time and strength which it is possible to obtain within the given constraints. However, experimental conditions which give short disintegration times also give weak tablets, and so a compromise solution, involving both dependent variables, must be sought.

It will be noted that Figs. 7.3 and 7.4 are plotted on identical axes, and hence can be superimposed. This gives a 'window' in which is contained all permissible combinations of disintegrant concentration and compression pressure which give tablets that comply with the imposed constraints of dependent and independent variables. This process is often facilitated by drawing these graphs on transparent sheets and superimposing them.

Calculations of the type above can be used to ascertain the maximum (or minimum) values of a dependent variable given certain constraints. A more usual

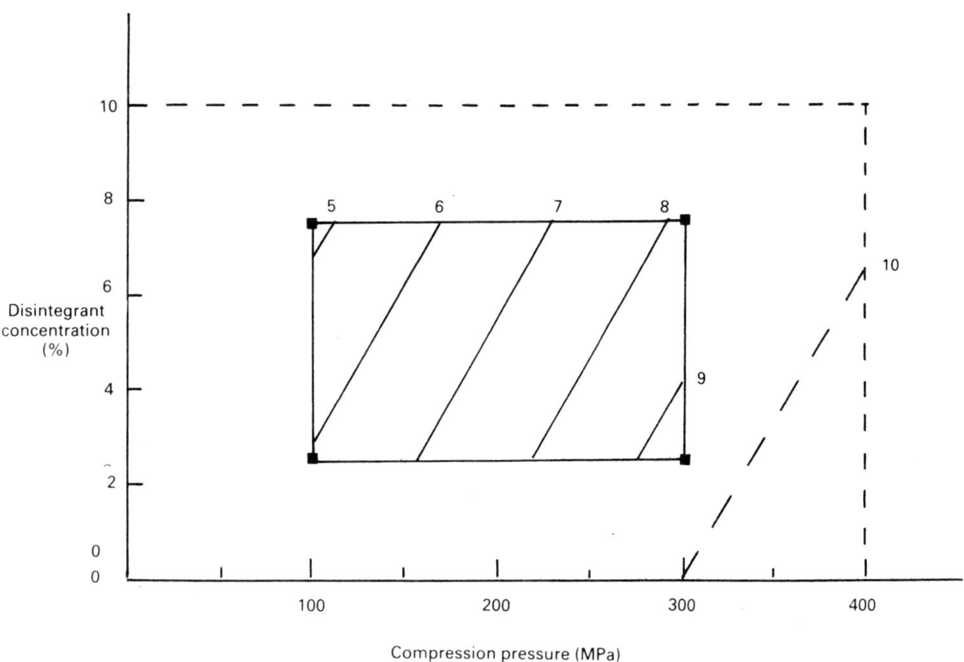

Fig. 7.4 — Contour plot of tablet strength derived from a two-factor, two-level factorial design.

problem is to specify ranges of values for the dependent variable and then attempt to ascertain the values of the independent variables needed to meet that specification.

As an example, suppose that it is required to produce tablets whose disintegration time does not exceed 600 seconds and whose breaking strength exceeds 6 kg, the process being subject to the same constraints as before. The so-called solution space is represented by the hatched portion of Fig. 7.6. Any combination of pressure and disintegrant concentration lying in this area should give tablets with the specified properties. Thus any of the following combinations, obtained without extrapolation, should suffice.

Disintegrant concentration (%)	Pressure range (MPa)
3	100–140
4	120–170
5	130–200
6	140–230
7	160–260

Extrapolation gives the following additional combinations.

8	170–290
9	190–350
10	200–350

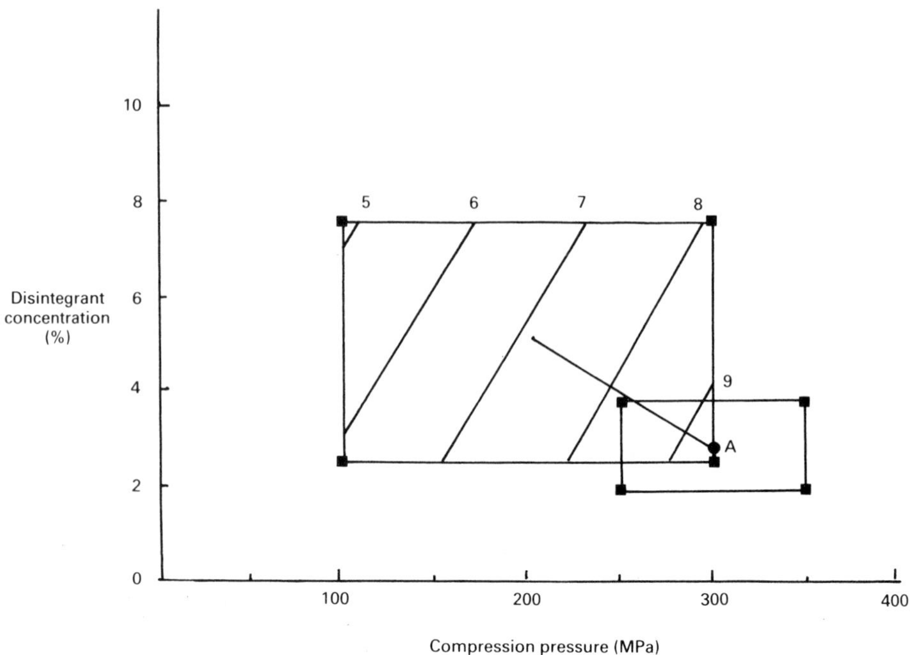

Fig. 7.5 — Procedure for determining the position of a second factorial design, using the path of steepest ascent method.

Which of the combinations is chosen will depend on other factors. For example, it would be best to avoid combinations which are at or near the constraints, or which give tablets whose properties are near the specification limits. Alternatively a cost criterion may be appropriate. This probably would not be applicable in the present case, but could apply when both independent variables are concentrations of two of the ingredients. In such circumstances, the cheapest combination would be chosen.

Of equal importance is the use of such diagrams to ascertain if specifications are feasible. For example, a specification that tablets should disintegrate in less than 300 seconds and have a strength of not less than 8 kg cannot be achieved, since there is no combination of disintegrant concentration and pressure which will yield such tablets while remaining within experimental constraints.

ANOVA WITH RESPONSE SURFACE METHODOLOGY

Such 2 × 2 designs can be made to answer the question 'How well do the equations describe the response surface?' This is done by replicating the points at the centre of the design. Thus in the example in Table 7.1, replicate experiments would be carried out at 200 MPa and 5% disintegrant. Such experiments provide a measure of the error of the system.

A further test of the experimental design is to select a combination of independent variables not studied hitherto and then make the tablets according to that

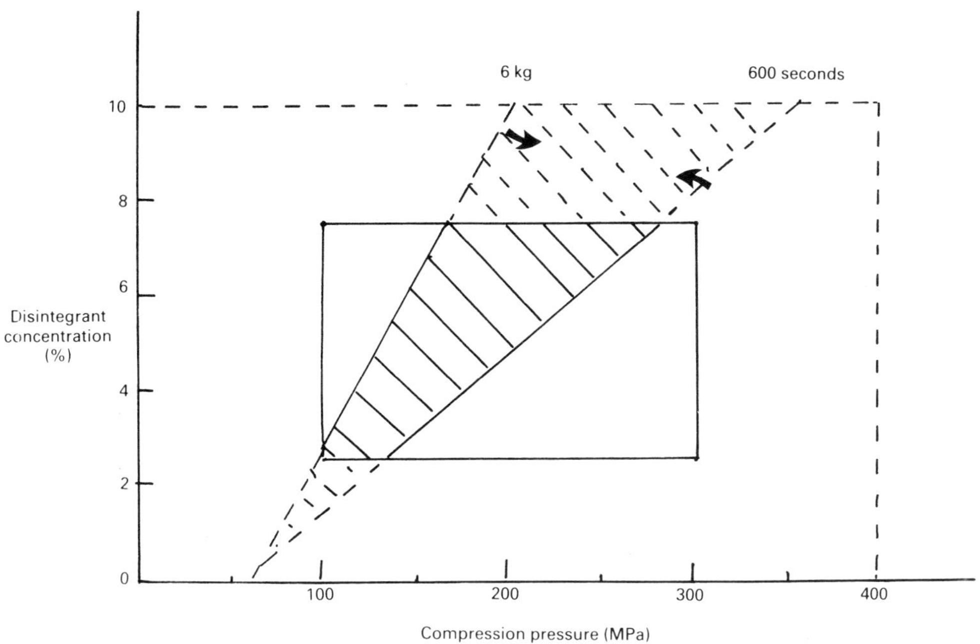

Fig. 7.6 — Combinations of compression pressure and disintegrant concentration which will give tablets of minimum strength 6 kg and maximum disintegration time of 600 seconds.

combination. Thus if a centre point experiment is to be carried out, then the regression equations (7.3) and (7.5) can be used to predict values of the dependent variables (tablets of disintegration time 586 seconds and of strength 7.2 kg). Confirming that such a formulation has the predicted properties is a test of the validity of the design and the equations derived from it.

OPTIMIZATION WHEN INTERACTION OCCURS BETWEEN THE INDEPENDENT VARIABLES

The previous example is one in which the two independent variables do not interact. It is essential to ascertain if this is the case, since if interaction does occur, then contour plots and any information derived from them will be significantly altered.

As before, this is best shown by an example. Consider the data shown in Table 7.2. This is identical to Table 7.1 except that one disintegration time has changed.

Plotting disintegration time against compression pressure gives Fig. 7.7, showing two lines which are not parallel, and hence indicating that an interaction is occurring.

In this case, the relationship between the two independent variables and disintegration time is given by an equation of the form:

$$Y_1 = B_0 + B_1X_1 + B_2X_2 + B_{12}X_1X_2 \tag{7.6}$$

Table 7.2 — The dependence of tablet strength and disintegration time on compression pressure and disintegrant concentration

Experiment	Compression pressure (MPa) (X_1)	Disintegrant concentration (%) (X_2)	Disintegration time (sec) (Y_1)	Tablet strength (kg) (Y_2)
—				
(1)	100	2.5	500	6.1
x_1	300	2.5	1070	9.4
x_2	100	7.5	140	4.9
x_1x_2	300	7.5	290	8.2

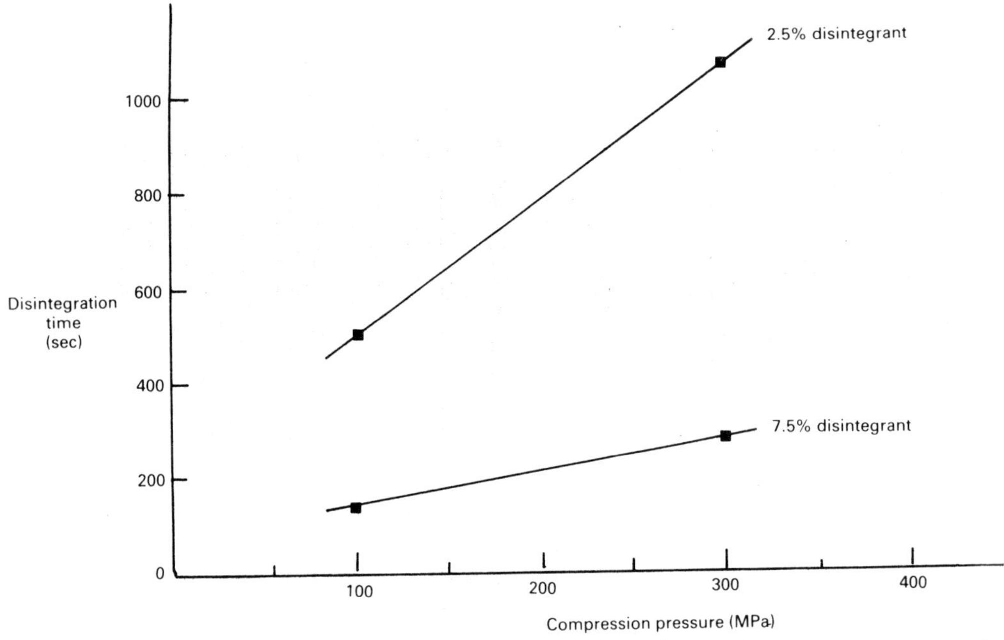

Fig. 7.7 — The relationship between compression pressure and tablet disintegration time, using two concentrations of disintegrant. Interaction occurs between the two independent variables.

This can be arranged to give:

$$X_1 = \frac{Y_1 - B_0 - B_2X_2}{B_1 + B_{12}X_2} \tag{7.7}$$

Solving equation (7.7) by multiple regression gives:

$$Y_1 = 290 + 3.90X_1 - 30.0X_2 - 0.42X_1X_2 \tag{7.8}$$

This in turn can be used to give Fig. 7.8. Note that a series of curved lines is now

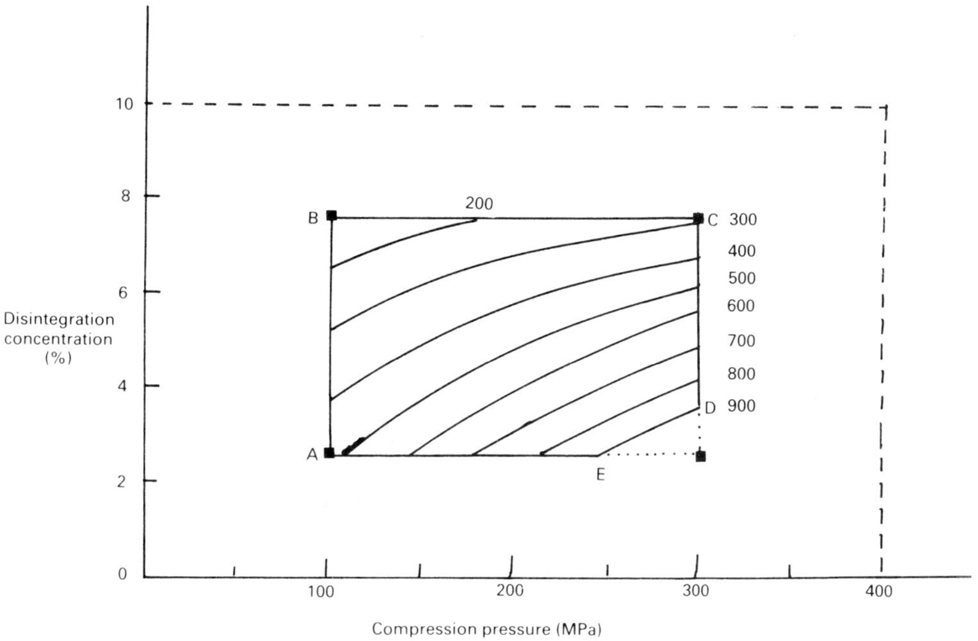

Fig. 7.8 — Contour plot of tablet disintegration time: interaction occurs between the two independent variables.

obtained. Superimposition of the graph involving tablet strength data as before gives a window in which all permissible solutions appear.

USE OF CODED DATA

In the discussion so far, the independent variables have been expressed in terms of their actual units, e.g. MPa, %. However, it is often easier, especially when considering more than two independent variables, to use coded values and express them in so-called 'experimental units'.

Thus in a two-factor, two-level design, the midpoint is termed (0, 0), and the four corners of the design (1, 1), (1, − 1), (− 1, − 1), and (− 1, 1) respectively (Fig. 7.9).

Using the data in Table 7.2, the midpoint is 200 MPa, 5% disintegrant. As the actual pressures used in the study are 100 and 300 MPa, it follows that 1 experimental unit for pressure is 100 MPa. By a similar argument, an experimental unit for disintegrant concentration is 2.5%. An advantage of this system is that it enables

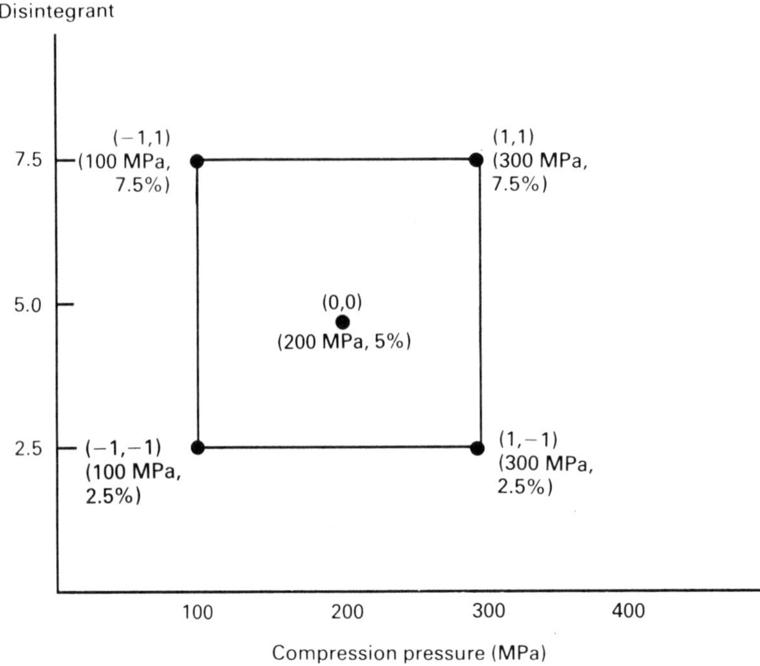

Fig. 7.9 — Two-factor, two-level design using coded data.

values of the dependent variable to be plotted against two or more independent variables on the same two-dimensional graph (Fig. 7.10).

SECOND-ORDER RELATIONSHIPS BETWEEN INDEPENDENT AND DEPENDENT VARIABLES

The discussion so far has dealt with situations where there is a linear relationship between independent and dependent variables, or where an interaction occurs between them.

In many cases, such a relationship sufficiently accurately reflects the actual situation, and in others, a first-order relationship is adequate to locate the approximate area in which the optimum is to be found. In the great majority of formulation problems, this is probably sufficient. If however it is necessary to find the position of the optimum with a greater degree of precision, or the actual value of the dependent variable at the optimum is required, then it is likely that a second-order relationship has to be used, perhaps after an approximate optimum has been located with a linear model.

In general terms, such a model has the form

$$Y = B_0 + B_1X_1 + B_2X_2 + B_{11}X_1^2 + B_{22}X_2^2 + B_{12}X_1X_2 \tag{7.9}$$

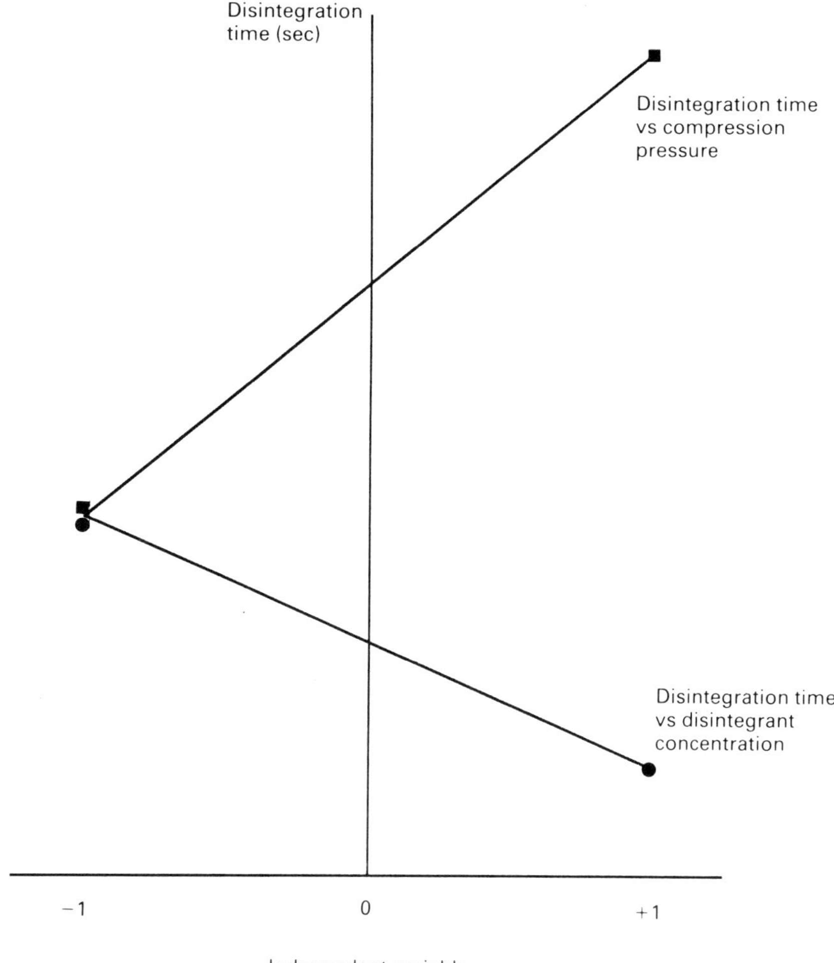

Fig. 7.10 — Tablet disintegration time plotted against disintegrant concentration (●) and compression pressure (■) using coded data.

This has six unknown coefficients, and hence a two-factor, two-level design provides insufficient points to solve six simultaneous equations. Additional experimental points are made available by use of what is known as a central composite design. This provides four extra points which have the coordinates ($X_1 = 0$, $X_2 = \pm 1.414$) and ($X_1 = \pm 1.414$, $X_2 = 0$), expressing values in terms of experimental units. The design is shown pictorially in Fig. 7.11.

If there were three independent variables, then the relevant model would be:

$$Y = B_0 + B_1X_1 + B_2X_2 + B_3X_3 + B_{11}X_1^2 + B_{22}X_2^2 + B_{33}X_3^2 + B_{12}X_1X_2 + \\ + B_{13}X_1X_3 + B_{23}X_2X_3 \qquad (7.10)$$

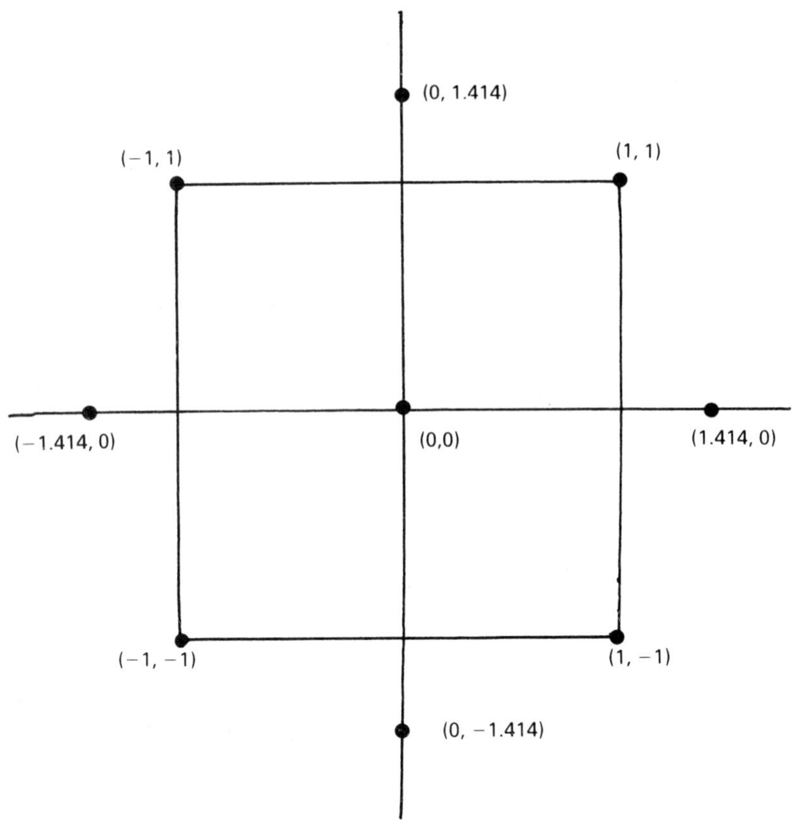

Fig. 7.11 — Two-factor central composite design.

This has ten unknown coefficients. Again, a central composite design will provide the necessary number of goups of experimental conditions, as shown in Fig. 7.12. In this case, the axial points are (0, 0, ± 1.68), (0, ± 1.68, 0) and (± 1.68, 0, 0).

The values of the additional axial points expressed in experimental units are chosen so that the whole design can be rotated around a central point. Thus all the points in Fig. 7.11 form a circle if the design is rotated around (0, 0). For Fig. 7.12, rotation provides a sphere.

If the relationships between the independent variables and the dependent variables are known (or suspected) to be non-linear, then it may be advantageous to assume a second-order relationship at the outset, and adopt a suitable experimental design. Thus, for example, suppose that it is suspected that the relationships between disintegrant concentration and compression pressure as independent variables, and tablet strength and disintegration time as dependent variables are all non-linear.

A suitable factorially-designed experiment is shown in Table 7.3, using three levels of both independent variables. Multiple regression relates each dependent variable to the two independent variables by equation (7.11).

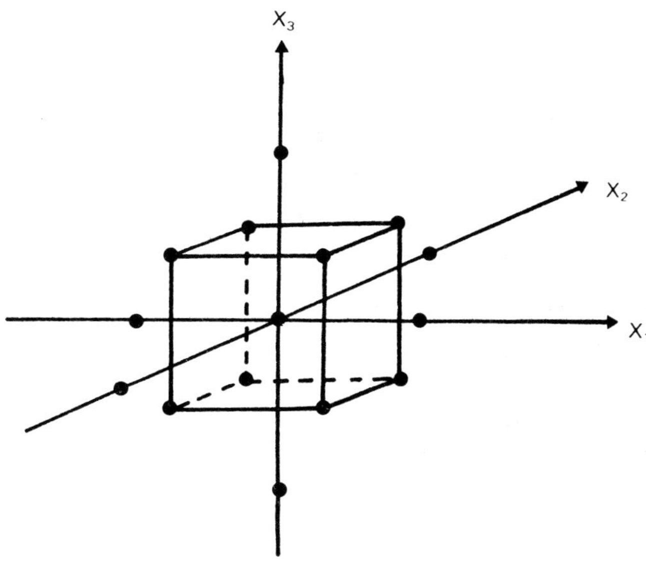

Fig. 7.12 — Three-factor central composite design.

Table 7.3 — The dependence of tablet strength and disintegration time on compression pressure and disintegrant concentration, using a three-level study

Experiment	Compression pressure (MPa)	Disintegrant concentration (%)	Disintegration time (sec)	Tablet strength (kg)
(00)	100	0	500	8.1
(10)	200	0	680	14.8
(20)	300	0	1020	16.4
(01)	100	5	220	11.4
(11)	200	5	400	19.3
(21)	300	5	720	23.2
(02)	100	10	70	2.0
(12)	200	10	80	8.2
(22)	300	10	270	11.8

$$Y = B_0 + B_1X_1 + B_{11}X_1^2 + B_2X_2 + B_{22}X_2^2 + B_{12}X_1X_2 \tag{7.11}$$

Rearrangement of equation (7.11) gives

$$B_{11}X_1^2 + (B_1 + B_{12}X_2)X_1 + (B_2X_2 + B_{22}X_2^2 + B_0 - Y) = 0 \tag{7.12}$$

Thus if the coefficients B_0, B_1, B_{11}, B_2, B_{22} and B_{12} are known, then the values of X_1 and X_2 needed to give a specified value of Y can be obtained by solving the quadratic equation (7.12). The solution is given in equation (7.13).

$$X_1 = -(B_1 + B_{12}X_2) \pm \frac{\sqrt{(B_1 + B_{12}X_2)^2 - 4B_{11}(B_2X_2 + B_{22}X_2^2 + B_0 - Y)}}{2B_{11}} \tag{7.13}$$

Multiple regression of the data in Table 7.3 gives the coefficients shown in Table 7.4.

Table 7.4 — Coefficients derived by multiple regression of data given in Table 7.3.

Coefficient	Dependent variable	
	Disintegration time	Tablet strength
B_0	433	−2.62
B_1	−0.37	0.124
B_{11}	0.008	−0.000195
B_2	−23.3	2.37
B_{22}	−0.40	−0.31
B_{12}	−0.16	0.00075

values of X_1, given X_2 and Y, the task being facilitated by use of a computer spreadsheet.

Thus for values of disintegration time increasing in multiples of 100 seconds, a series of contours is obtained (Fig. 7.13). Constraints can be applied as before, and so combinations of independent variables which give specified values of the dependent variable can be obtained.

A similar treatment can be applied to the tablet strength data (Fig. 7.14) and the two sets of contour plots can be combined, also as described earlier, to locate the area of the optimum solution.

As the number of independent variables is increased, so does the complexity of the second-order model used to describe them. A system using five independent variables was described by Schwartz *et al.* in 1973. These workers looked at five formulation parameters relevant to a tablet formulation. These were diluent composition, compression pressure, disintegrant content, granulating agent content and lubricant content.

They also measured eight dependent variables (disintegration time, hardness, dissolution rate, friability, weight, thickness, porosity and mean pore diameter).

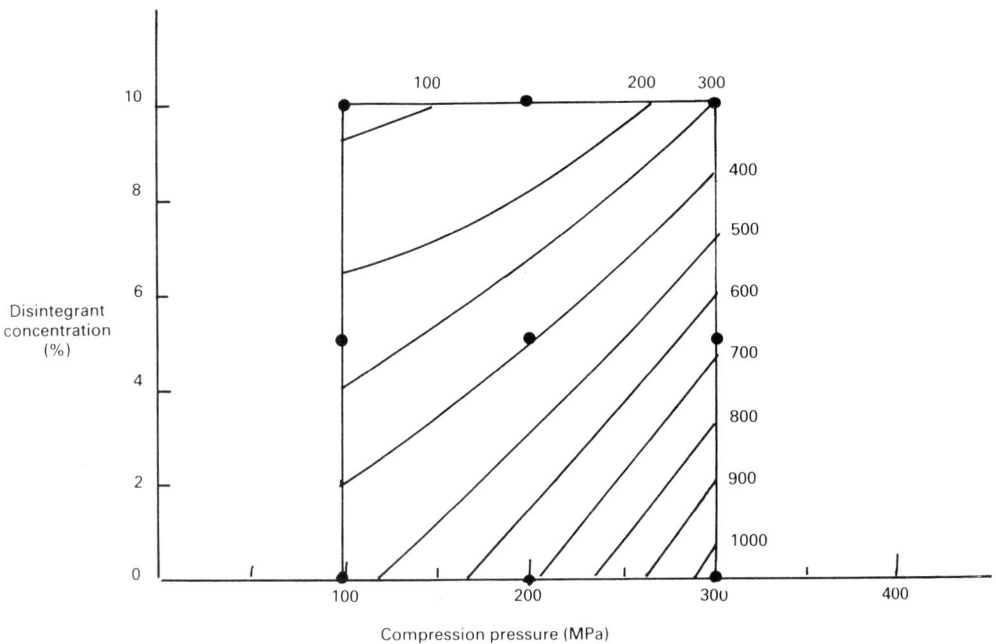

Fig. 7.13 — Contour plot of tablet disintegration time derived from a two-factor, three-level factorial design with interaction between the factors.

Each dependent variable was fitted into an equation of all independent variables as shown in equation (7.14):

$$Y_1 = B_0 + B_1X_1 + \ldots + B_5X_5 + B_{11}X_1^2 + \ldots + B_{55}X_5^2 + B_{12}X_1X_2 + \ldots + \\ + B_{45}X_4X_5 \qquad (7.14)$$

This contains twenty-one unknown coefficients and hence an experimental design must be chosen which will provide sufficient data points for these to be calculated.

The design used by Schwartz and co-workers is shown in Table 7.5. Experimental units are used throughout.

The first sixteen experiments represent a half-factorial design for five factors at two levels. A full factorial design would comprise thirty-two experiments, and so the reduction to sixteen involves some confounding. However, none of the two-way interactions are confounded with main effects nor with each other, but three-way interactions are considerably confounded, e.g. $X_1X_2X_3$ with X_4X_5.

The remainder of the twenty-seven experiments are needed to provide sufficient experimental points to satisfy the number of coefficients, and also to achieve symmetry. Thus for each factor three additional levels were selected. Zero repre-

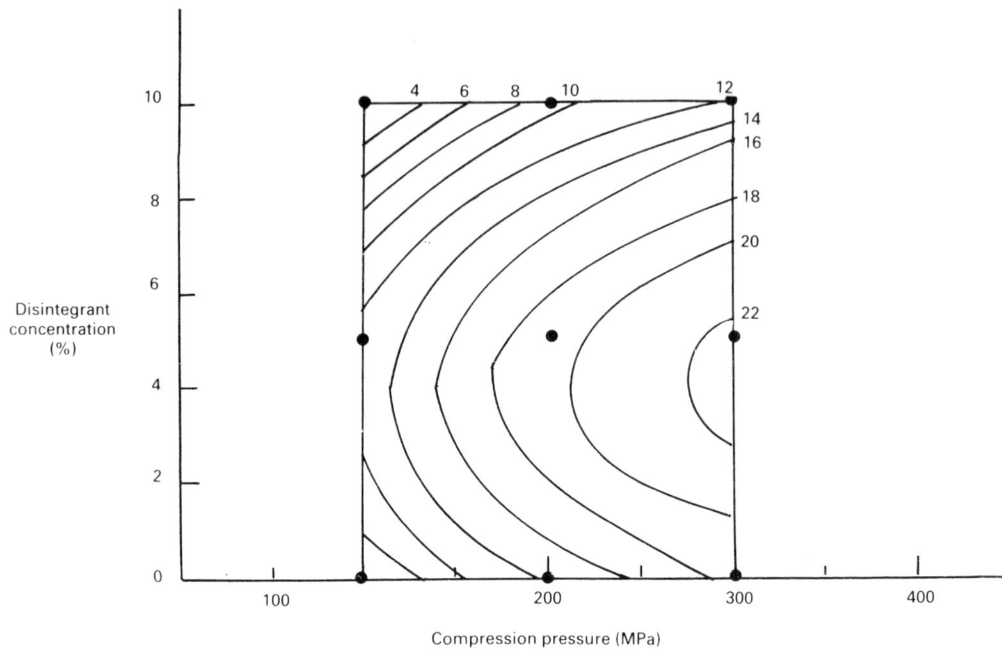

Fig. 7.14 — Contour plot of tablet strength derived from a two-factor, three-level factorial design with interaction between the factors.

sents the midpoint of each factor of the design, and + 1.547 and − 1.547 the extreme values for each variable. Experiment 27 represents the midpoint of the whole experimental design, with all five factors set to zero level.

In this study, factor X_5 is lubricant content in milligrams and one experimental unit represents 0.5 mg of lubricant. Therefore, the five levels of lubricant, which in experimental unit terms are − 1.547, − 1, 0, + 1 and + 1.547, are, when expressed in physical units, 0.2, 0.5, 1.0, 1.5 and 1.8 mg respectively.

COMPUTER SOFTWARE PACKAGES FOR OPTIMIZATION

Several software packages are now available to give assistance in the areas of experimental design and optimization. An example is the RS/Discover suite of programs from BBN Software Products.

The menu-driven program invites the user to specify the independent variables, together with their units, the ranges of values of the variables, the required degree of precision, and to indicate if the value of a given variable can be readily altered.

The program then produces a worksheet which gives the design of the experiment (full factorial, central composite, partial factorial, etc.) and the values of the independent variables for each experiment. The experiments are normally given in random order except where a particular experimental variable cannot be readily

Table 7.5 — Experimental design for five independent variables (Schwartz *et al.* 1973)

Experiment	Factor level in experimental units X_1	X_2	X_3	X_4	X_5
1	−1	−1	−1	−1	1
2	1	−1	−1	−1	−1
3	−1	1	−1	−1	−1
4	1	1	−1	−1	1
5	−1	−1	1	−1	−1
6	1	−1	1	−1	1
7	−1	1	1	−1	1
8	1	1	1	−1	−1
9	−1	−1	−1	1	−1
10	1	−1	−1	1	1
11	−1	1	−1	1	1
12	1	1	−1	1	−1
13	−1	−1	1	1	1
14	1	−1	1	1	−1
15	−1	1	1	1	−1
16	1	1	1	1	1
17	−1.547	0	0	0	0
18	1.547	0	0	0	0
19	0	−1.547	0	0	0
20	0	1.547	0	0	0
21	0	0	−1,547	0	0
22	0	0	1.547	0	0
23	0	0	0	−1.547	0
24	0	0	0	1.547	0
25	0	0	0	0	−1.547
26	0	0	0	0	1.547
27	0	0	0	0	0

changed in value. In such cases, experiments are grouped so that time used for changing the independent variable is minimized.

After the experiments have been carried out, the responses can be added to the worksheet. Data can then be analysed, fitted to models, and contour plots and response surfaces produced.

Applications of this system have been given by McGurk *et al.* (1989) and Jones *et al.* (1989).

REFERENCES

Jones, S. P., Sandhu, G. and Lendrem, D. W. (1989) Enteric coating: formulation, optimisation and sampling technique, *J. Pharm. Pharmacol.* **41** 130P.

McGurk, J. G., Storey, R. and Lendrem, D. W. (1989) *Computer-aided process optimisation*, *J. Pharm. Pharmacol.* **41** 128P.

Schwartz, J. B., Flamholz, J. R. and Press, R. H. (1973) Computer optimisation of pharmaceutical formulations, *J. Pharm. Sci.* **62** 1165–1170 and 1518–1519.

ADDITIONAL READING

Bolton, S. (1984) *Pharmaceutical Statistics*, Marcel Dekker.

Burley, D. M. (1974) *Studies in Optimisation*, Intertext.

Jones, B. A. (1987) Design lab experiments to assure product quality, *Research & Development*, December 1987.

Formulating with statistics (1990), *Pharm. J.* **243** 88.

8

Model-independent optimization

OPTIMIZATION BY SIMPLEX SEARCH

The simplex search method is an optimization procedure which adopts an empirical approach rather than making assumptions about the response surface. The results of previous experiments are used to define the experimental conditions of subsequent experiments in an attempt to find the optimal response. The optimum is approached by moving away from the low values of the response.

The name simplex derives from the shape of the geometric figure which moves across the response surface. It is defined by a number of vertices equal to one more than the number of variables in the space. Thus a simplex of two variables is a triangle.

The basis of the method is most readily grasped in the case where there are two independent variables, X_1 and X_2. The simplex is constructed by selecting three combinations of these two variables (A, B, C). The three experiments represented in Fig. 8.1 are carried out and the response measured in each case (R_A, R_B, R_C). The worst response is identified (for example R_A), and the values of the independent variables for the next experiment *D* are chosen by moving away from point A. This is achieved by reflecting the triangle ABC about the BC axis. Hence AP = DP. The experiment at point D is performed and the response R_D compared with the responses at points A, B and C.

The next move depends on the relative values of the four responses.

If R_D is greater than R_A, R_B or R_C, then it is worthwhile proceding further along the AD axis, and the next point E is located at (P + 2AP) along this line. This procedure is termed expansion.

If R_D is greater than R_B, but less than R_C, then vertex D is retained, and the next point, F, is located by moving away from B, reflecting triangle BCD about axis CD.

If the response at D is lower than that at B and C but greater than that at A, the procedure is to locate the next experiment (G) along the AD axis at (P + 0.5AP). On the other hand, if R_D is lower than R_A, R_B or R_C, then point H is located along the same axis at (P − 0.5AP). These procedures are known as contractions.

The overall position is given in Table 8.1 and summarized in Fig. 8.2

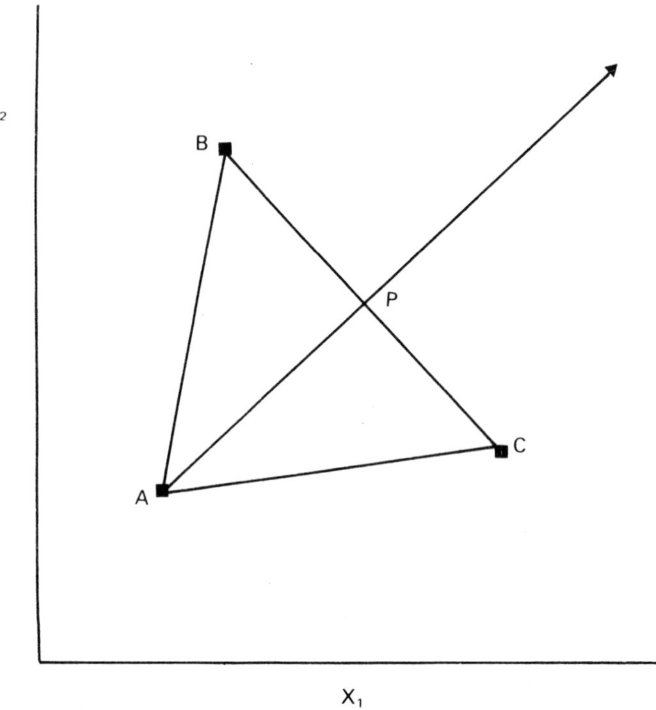

Fig. 8.1 — The first stage in optimization by simplex search.

Table 8.1 — Procedure to determine course of action after responses have been obtained at points A, B, C and D

Relative value of response	Course of action
$R_D > R_A, > R_B, > R_C$	Expand further along line APD.
$R_D > R_A, > R_B, < R_C$	Reflect triangle BCD about CD axis.
$R_D > R_A, < R_B, < R_C$	Contract along line PD.
$R_D < R_A, < R_B, < R_C$	Contract along line AP.

An example of how the simplex approach can be applied is provided by the work of Gould and Goodman (1983). These workers used the technique to determine the blend of ethanol, propylene glycol and water in which caffeine showed maximum solubility.

The vertices, the percentages of ethanol and propylene glycol and the solubility of caffeine in that blend are shown in Table 8.2.

Of the three chosen combinations of the solvents, point 3 gives the lowest

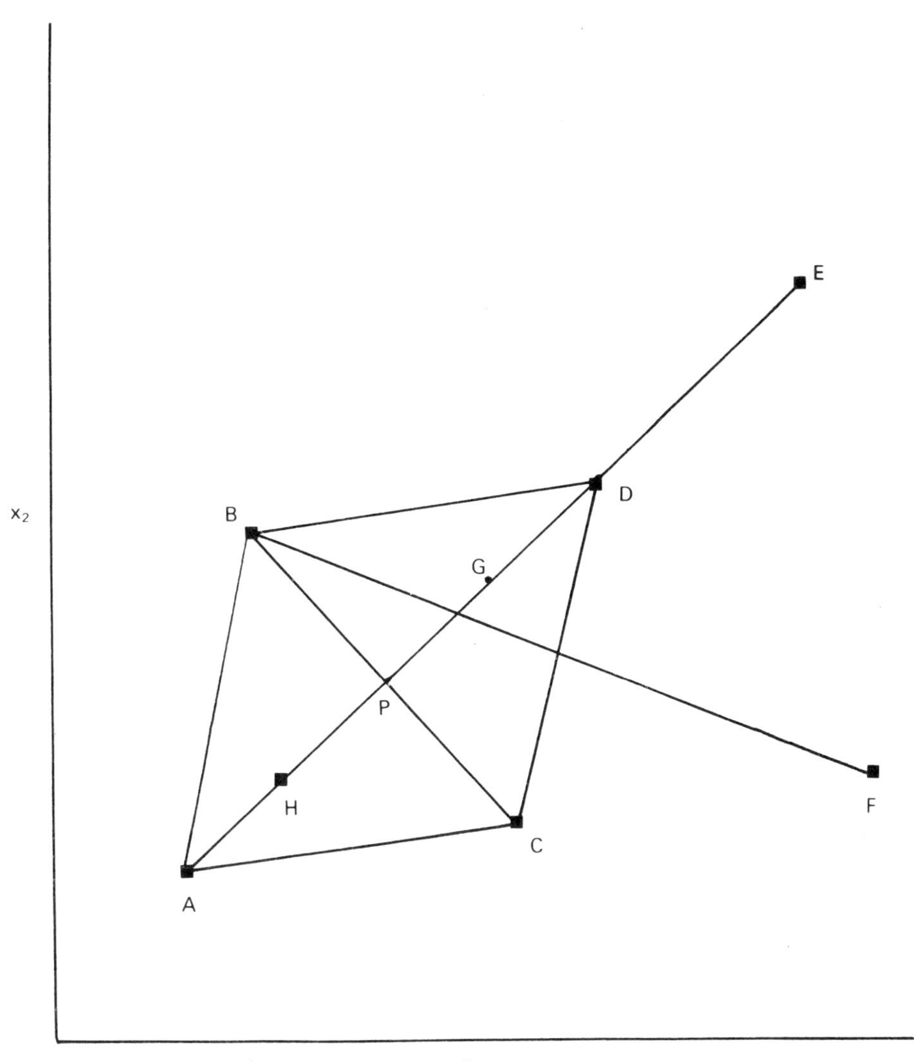

Fig. 8.2 — Subsequent stages in optimization by simplex search.

solubility, and hence point 4 is located by reflection about the 1–2 axis. Solubility at point 4 is higher than that at points 2 and 3, and so expansion along the 1–4 axis to point 5 is indicated. This gives the lowest solubility of all, and so further progression along this line is not worthwhile. Consideration thus returns to points 1, 2 and 4. Of these, point 1 is the lowest, and so reflection from that point about the 2–4 axis gives point 6. The triangle 2, 4, 6 is now considered. Point 2 is the lowest of these three points, so reflection now occurs about the 4–6 axis to give point 7. Solubility at point 7 is lower than that at both points 4 and 6, but higher than at point 2, so contraction to give points 8 and 9 is carried out. The last two points are virtually the same, indicating

Table 8.2 — Vertices, solvent blends and solubilities of caffeine in those solvent blends (Gould and Goodman 1983)

Vertex	Ethanol (%v/v)	Propylene glycol (%v/v)	Solubility ($mgml^{-1}$)	Vertices retained	Vertex rejected	Process
1	0	40	24.0	—	—	—
2	20	0	26.2	—	—	—
3	0	0	17.2	—	—	—
4	20	40	44.9	1, 2	3	Reflection
5	30	60	17.5	1, 2	3	Expansion
6	40	0	52.4	2, 4	1	Reflection
7	40	40	36.7	4, 6	2	Reflection
8	35	30	52.9	4, 6	7	Contraction
9	29	28	53.0	4, 6	8	Contraction

that a maximum is nearby. The precise point of the maximum could be found by further experiments if this was considered worthwhile.

A more complex example is given by the work of Shek *et al.* (1980) who attempted to optimize a capsule formulation by a simplex approach. In this case, there were four independent variables, namely the concentrations of drug, disintegrant and lubricant, and the total capsule weight. Thus the simplex for this design is a pentagon. Three of the four independent variables have units of percentage concentration, but the fourth has a different unit, weight. All four variables must be put on the same unitary basis, and this is achieved by normalization. Furthermore it is also desirable to place upper and lower limits on the values of each independent variable, so that possible combinations of independent variables are composed of achievable values. This last point is the equivalent of setting constraints in a model-dependent optimization procedure.

Normalization is carried out by using equation (8.1)

$$N = \frac{X - L}{H - L} \times 100\% \tag{8.1}$$

where N is the normalized value and X is the original uncorrected value of that variable, and L and H are the lowest and highest values of that factor which are likely to be of interest.

Thus, for example, Shek *et al.* (1980) state that the lowest and highest values of capsule fill weight would be 100 and 400 mg respectively, these values presumably being derived from the sizes of available capsule shells and/or filling equipment. Thus a capsule weight of 200 mg, when normalized, would become

$$\frac{200 - 100}{400 - 100} \times 100\% = 33.3\%$$

A further point which must be considered is whether there is more than one dependent variable whose magnitude is considered important. Shek *et al.* (1980) decided that there were three responses of interest, namely rate of packing down of the powder (R_1), % dissolved at 30 minutes (R_2) and % dissolved at 8 minutes (R_3). Though it would be possible to maximize each of these responses, the values of the independent variables for each individual response would be different. Hence it is necessary to combine these responses and to arrange them in order of importance. Then each can be given an appropriate weighting which reflects the importance of that factor to the overall success of the formulation. Thus if R_t is the total response, then

$$R_t = 0.5R_1 + 0.4R_2 + 0.1R_3 \tag{8.2}$$

Since the responses have different units, and a range of desired values can be defined, they are normalized using the formula given in equation (8.1).

It must be stressed that the coefficients in equation (8.2) are arbitrary and are chosen by the experimenter. If all three responses are considered equally important, each coefficient would equal 0.33.

Each of the responses is measured and the total response calculated according to equation (8.2).

Thus the calculated value R_t has the units of (%) and the precise optimum combination of responses will therefore have a value of 100. It is unlikely that this would ever be achieved, and so experimentation can be reduced by specifying a lower but acceptably high value for R_t, for example 80%.

The simplex approach can be regarded as a step-by-step process of achieving the optimum. It must be conceded that many steps may be needed before that optimum is reached. For example, Gould and Goodman (1983) carried out nine experiments, and Shek *et al.* (1980) forty-five experiments before a satisfactory optimum was achieved. However, a willingness to settle for less than the precise optimum greatly reduces the number of experiments. Thus if Gould and Goodman (1983), rather than searching for the maximum, had looked for a solvent mixture in which caffeine was soluble in excess of 50 $\mathrm{mg\,ml^{-1}}$, then this would have been discovered in six experiments.

Though the numbers of experiments may seem high, it is worth noting that a full three-level factorial design, using four factors, would necessitate eighty-one experiments. Even then, derivation of the model might show that the optimum lay outside the chosen ranges of values for the independent variables. This would necessitate extrapolation or repeating the design over other ranges.

However, if one is willing to accept less than the optimal response, then it must also be accepted that there will be a number of combinations of experimental conditions which will give this response. Thus, for example, Gould and Goodman (1983) found a solvent blend in which the solubility of caffeine exceeded 50 $\mathrm{mg\,ml^{-1}}$ in their sixth experiment. Even so, this may not be the 'best' blend of ethanol, propylene glycol and water which gives this solubility. An obvious further consideration is cost. If the three components of the mixture differ in cost, as seems likely in this case, then it is sensible to select the cheapest combination which gives the required effect. It may therefore be useful to combine the model-independent simplex approach with a model-dependent technique. Using data derived in the simplex search, regression analysis is carried out followed by mapping of the response surface. This combined approach was used by Shek *et al.* (1980), and the reader is referred to their paper for further details.

REFERENCES

Gould, P. L. and Goodman, M. (1983) Simplex search in the optimisation of the solubility of caffeine in parenteral cosolvent systems, *J. Pharm. Pharmacol.* **35** 3P.

Shek, E., Ghani, M. and Jones, R. E. (1980) Simplex search in optimisation of capsule formulation, *J. Pharm. Sci.*, **69** 1135–1141.

ADDITIONAL READING

Gould, P. L. (1984) Optimisation methods for the development of dosage forms, *Int. J. Pharm. Tech. & Prod. Mfr.* **5**(1) 19–24.

Appendix 1: Computer programs in BASIC and MINITAB commands

By removing the drudgery of repetitive arithmetical calculations, the availability of computers has made the techniques of experimental design much more accessible. A major comprehensive statistical package such as MINITAB (Minitab Inc., State College, Pennsylvania, USA) provides most of the necessary mathematical features, but even a PC provides considerable assistance.

In this appendix are programs, written in BASIC and suitable for use on a PC, which the authors have found to be useful. Each program is demonstrated using data taken from the text. Reference is also made to the chapters of the text where these programs are most useful.

If MINITAB commands are available for the same purpose, then those are also given.

Additional reading

Ryan, B. F., Joiner, B. L. and Ryan, T. A. (1985) Minitab Handbook, PWS Publishers.

1.1 CALCULATION OF MEAN, STANDARD DEVIATION, ETC.

This BASIC program is used to calculate the sum, mean, sum of squares, variance, standard deviation, coefficient of variation and standard error of the mean of a set of numbers. The data are then reproduced in standard form. It can be used in Chapters 2, 3, 4, 5, 6.

Operation (Tables 6.1 and 6.2, acid value data).

Display	Response
Enter number of elements	
	5

Enter value of elements	
	1.0
	1.4
	1.2
	1.5
	1.3
Number of elements	5
Mean	1.28
Variance	0.037
Standard deviation	0.192
Coefficient of variation	15%
Standard error of the mean	0.086

Raw data	Standardized data
1.0	1.456
1.4	0.625
1.2	−0.417
1.5	1.144
1.3	0.104

Hard copy required? (y/n)
Another data set? (y/n)

```
100 REM Program 'Mean'
110 REM This program calculates the sum, mean, sum of squares, variance, standar
d deviation, standard error of mean, and reproduces the data in standard form.
120 DIM X(200)
130 PRINT "Enter number of elements"
140 INPUT N
150 LET SUM = 0
160 LET SDXS = 0
170 REM enter data
180 FOR I=1 TO N
190 PRINT "Enter value of element"
200 INPUT X(I)
210 SUM = SUM + X(I)
220 NEXT I
230 LET MEAN = SUM/N
240 REM calculate total sum of squares of deviations
250 FOR J=1 TO N
260 DX(J)=X(J)-MEAN
270 DXS(J)=DX(J)^2
280 SDXS=SDXS+DXS(J)
290 NEXT J
300 REM calculate variance etc
310 VAR=SDXS/(N-1)
320 VAR=INT(VAR*1000+.5)/1000
330 SD=SQR(VAR)
340 SD=INT(SD*1000+.5)/1000
350 SE=SD/(SQR(N))
360 SE=INT(SE*1000+.5)/1000
370 CV=SD*100/MEAN
380 CV=INT(CV*1000+.5)/1000
390 PRINT "Number of elements" TAB(40)N
```

```
400 PRINT "Mean" TAB(40)INT(MEAN*1000+.5)/1000
410 PRINT "Variance"TAB(40)VAR
420 PRINT "Standard deviation"TAB(40)SD
430 PRINT "Coefficient of variation"TAB(40)CV"%"
440 PRINT "Standard error of the mean"TAB(40)SE
450 REM standardised data
460 PRINT TAB(20)"Raw data";TAB(40)"Standardised data"
470 FOR K = 1 TO N
480 DX(K)=X(K)-MEAN
490 EX(K)=DX(K)/SD
500 EX(K)=INT(EX(K)*1000+.5)/1000
510 PRINT TAB(20)X(K);TAB(40)EX(K)
520 NEXT K
530 REM hard copy of data and results
540 PRINT "Hard copy required?(y/n)
550 INPUT Q2$
560 IF Q2$="n" GOTO 720
570 LPRINT
580 LPRINT "Number of elements" TAB(40)N
590 LPRINT"Mean"TAB(40)MEAN
600 LPRINT "Variance" TAB(40)VAR
610 LPRINT "Standard deviation"TAB(40)SD
620 LPRINT "Coefficient of variation"TAB(40)CV"%"
630 LPRINT "Standard error of mean"TAB(40)SE
640 LPRINT
650 LPRINT TAB(20)"Raw data";TAB(40) "Standardised data"
660 FOR K=1 TO N
670 LPRINT TAB(20)X(K);TAB(40)DX(K)/SD
680 NEXT K
690 LPRINT
700 LPRINT
710 LPRINT
720 PRINT "Another data set?(y/n)
730 INPUT Q1$
740 IF Q1$="n" GOTO 770
750 CLS
760 GOTO 130
770 END
```

MINITAB command

DESCRIBE

This command gives the following statistics: number of elements, mean, median, trimmed mean, standard deviation, standard error of the mean, minimum, maximum, first and third quartiles.

1.2 LINEAR REGRESSION

This program fits pairs of data (*X* and *Y*) into an equation of the form

$$Y=b_0+b_1X$$

It calculates the coefficient of *X* (i.e. b_1), the standard error of that coefficient and the *t* value. It also calculates the intercept (b_0), and its standard error and *t* value. The correlation coefficient and the standard error of the estimate are also calculated. Finally, if requested, a table is produced of the observed values of *Y* and calculated values of *Y*.

Operation (using data from Table 5.1).

Display	Response
Type in title	
	Viscosity
Type in the number of pairs of data, and press RETURN	
	5
Type data into table, pressing RETURN after each entry	
	12.3
	4.83
	18.5
	6.32
	24.6
	7.50
	30.8
	9.66
	36.9
	11.9
Which pair needs changing? Type 0 if all correct.	
	0
Summary	
Viscosity	
Number of data points	5
Coefficient	0.284
Standard error of coefficient	0.022
t value for coefficient	12.909
Intercept	1.05
Standard error of intercept	1046
t value for intercept	0.001
Correlation coefficient	0.991
Standard error of estimate	0.437
Do you want a table of observed and calculated results? Type y/n	
	y

Y_{obs}	Y_{calc}
4.83	4.5432
6.32	6.304
7.50	8.0364
9.66	9.7972
11.9	11.5296

```
100 REM linear regression and correlation
110 PRINT"Type in title"
120 INPUT J$
130 PRINT J$
140 PRINT
150 PRINT"Type in the number of pairs of data and press RETURN"
160 INPUT A
170 DIM X(A):DIM Y(A):DIM Z(A)
180 PRINT
190 PRINT"Type data into the table, pressing RETURN after each entry"
200 PRINT
210 FOR J=1 TO A
220 INPUT X(J)
230 INPUT Y(J)
240 NEXT J
250 PRINT TAB(10)"X";TAB(18)"Y"
260 FOR J=1 TO A
270 PRINT TAB(8)X(J);TAB(16)Y(J)
280 NEXT J
290 PRINT
300 PRINT"Which pair needs changing? Type 0 if all correct.
310 INPUT CH:IF CH=0 THEN 360
320 PRINT"New values of "CH:
330 INPUT X(CH)
340 INPUT Y(CH)
350 GOTO 250
360 FOR J=1 TO A
370 SX=SX+X(J):REM sx=sx+x(J)
380 SY=SY+Y(J):REM sy=sum of y
390 SXSQ=SXSQ+(X(J)*X(J)):REM sxsq=sum of x squared
400 SYSQ=SYSQ+(Y(J)*Y(J)):REM sysq=sum of y squared
410 SXY=SXY+(X(J)*Y(J)):REM sxy=sum of products of x and y
420 NEXT J
430 SLOPE = (SXY-(SX*SY/A))/(SXSQ-(SX*SX/A))
440 SLOPE=INT(SLOPE*1000+.5)/1000
450 INTER=((SY/A)-(SLOPE*SX/A))
460 INTER=INT(INTER*1000+.5)/1000
470 XSS=SXSQ-SX*SX/A
480 YSS=SYSQ-SY*SY/A
490 SP=SXY-SX*SY/A
500 P=XSS*YSS
510 R=SP/SQR(P)
520 R=INT(R*1000+.5)/1000
530 SEE=SQR((YSS-SLOPE*SLOPE*XSS)/(A-2))
540 SEE=INT(SEE*1000+.5)/1000
550 SEC=SQR(SEE*SEE/XSS)
560 SEC=INT(SEC*1000+.5)/1000
570 SEI=SQR(SEE*SEE*((1/A)+(SX*SX/A*A*XSS)))
580 SEI=INT(SEI*1000+.5)/1000
590 TCOEFF=INT(SLOPE*1000/SEC+.5)/1000
600 TINTER=INT(INTER*1000/SEI+.5)/1000
610 CLS
620 PRINT"Summary"
630 PRINT
640 PRINT J$
650 PRINT
660 PRINT "Number of data points";TAB(50)A
670 PRINT
680 PRINT "Coefficient";TAB(50)SLOPE
690 PRINT "Standard error of coefficient";TAB(50)SEC
700 PRINT "t value for coefficient";TAB(50)TCOEFF
710 PRINT
720 PRINT "Intercept";TAB(50)INTER
730 PRINT"Standard error of intercept";TAB(50)SEI
740 PRINT "t value for intercept";TAB(50)TINTER
750 PRINT
760 PRINT "Correlation coefficient";TAB(50)R
770 PRINT
780 PRINT "Standard error of estimate";TAB(50)SEE
790 PRINT
800 PRINT "Do you want a table of observed and calculated results? Type y/n"
810 INPUT QI$
820 IF QI$="n" GOTO 890
830 PRINT
840 PRINT TAB(15) "Obs Y";TAB(30)"Calc Y"
```

```
850 FOR C=1 TO A
860 Z(C)=INTER+SLOPE*X(C)
870 PRINT TAB(15)Y(C);TAB(30)Z(C)
880 NEXT C
890 END
```

MINITAB command

REGRESS c2 on **1** predictor **c1**. (This assumes that values of X are in column 1 and values of Y in column 2).

This command gives the regression equation, the standard deviations of the coefficients, the t-ratios, and an analysis of variance.

1.3 PARABOLIC CURVE FIT

This program fits data to an equation of the form

$$Y=A+BX+CX^2$$

It calculates the values of the coefficients A, B and C, their confidence limits, the confidence limit of the estimate and the correlation coefficient.

It finally presents a table of X, the observed value of Y and the value of Y, calculated using the regression equation.

Operation (using data from Table 5.1).

Display	Response
Parabolic curve fit Type in title	
	Viscosity
Type in number of pairs of results and press RETURN	
	5
Type results into table, pressing RETURN after each entry	
	12.3 4.83 18.5 6.32 24.6 7.50 30.8 9.66 36.9 11.9

Number	X	Y
1	12.3	4.83
2	18.5	6.32

3	24.6	7.5
4	30.8	9.66
5	36.9	11.9

Which pair needs changing?
Type 0 if all correct.

0

Summary

The intercept is	3.677
The coefficient of X is	0.038
The coefficient of X squared is	0.005
The variance is	0.081
The confidence limit of the estimate is	0.201
The confidence limit of the intercept is	6.354
The confidence limit of the coefficient of *X* is	0.071
The confidence limit of the coefficient of *X* squared is	0.001
The correlation coefficient is	0.999
The *F* value is	393.039

X	Y_{obs}	Y_{calc}
12.3	4.83	4.907
18.5	6.32	6.108
24.6	7.50	7.671
30.8	9.66	9.645
36.9	11.90	11.969

```
100 PRINT "Parabolic curve fit"
110 PRINT "Type in title"
120 INPUT T$
130 PRINT "Type in number of pairs of results and press RETURN"
140 INPUT N
150 DIM X(N):DIM Y(N):DIM Z(N):DIM E(N):DIM F(N):DIM G(N):DIM D(N)
160 PRINT"Type results into table, pressing RETURN after each entry"
170 FOR I=1 TO N
180 INPUT X(I):INPUT Y(I)
190 NEXT I
200 PRINT TAB(10)"Number";TAB(20)"X";TAB(30)"Y"
210 FOR I=1 TO N
220 PRINT TAB(10)I;TAB(20)X(I);TAB(30)Y(I)
```

```
230 NEXT I
240 PRINT "Which pair needs changing? Type 0 if all correct"
250 INPUT CH
260 IF CH=0 THEN 300
270 PRINT "New values of "CH
280 INPUT X(CH):Y(CH)
290 GOTO 240
300 FOR I=1 TO N
310 SX=SX+X(I) :REM sx=sum of x
320 SY=SY+Y(I) :REM sy=sum of y
330 SXSQ=SXSQ+X(I)^2  :REM sxsq=sum of x squared
340 SYSQ=SYSQ+Y(I)^2 :REM sysq=sum of y squared
350 XYSUM=XYSUM+ X(I)*Y(I)  :REM xysum =sum of products of x and y
360 SCX=SCX+X(I)^3 :REM scx=sum of cubes of x
370 SFX=SFX+X(I)^4 :REM sfx=sum of x to fourth power
380 NEXT I
390 FOR I=1 TO N
400 D(I)=X(I)*X(I)
410 G(I)=D(I)*Y(I)
420 GB=GB +G(I):REM gb=sum of xsquared times y
430 NEXT I
440 H=(N*SXSQ-SX*SX)*(N*GB-SXSQ*SY)
450 K=(N*SCX-SX*SXSQ)*(N*XYSUM-SX*SY)
460 L=(N*SXSQ-SX*SX)*(N*SFX-SXSQ*SXSQ)-(N*SCX-SX*SXSQ)^2
470 M=(H-K)/L:REM m=coefficient of X squared
480 P=((N*XYSUM-SX*SY)-M*(N*SCX-SX*SXSQ))/(N*SXSQ-SX*SX):REM P=coefficient of X
490 S=(SY-M*SXSQ-P*SX)/N:REM s=intercept
500 MX=SX/N:REM mx=mean of x
510 MY=SY/N:REM my=mean of y
520 SK=SYSQ-SY*SY/N
530 SL=SXSQ-SX*SX/N
540 ST=SFX-SXSQ*SXSQ/N
550 SM=XYSUM-SX*SY/N
560 SN=GB-SY*SXSQ/N
570 SP=SCX-SX*SXSQ/N
580 FOR I=1 TO N
590 Z(I)=S+P*X(I)+M*X(I)*X(I)
600 E(I)=(Z(I)-MY)^2
610 SA=SA+E(I)
620 F(I)=(Y(I)-MY)^2
630 SF=SF+F(I)
640 T(I)=(Z(I)-Y(I))^2
650 V=V+T(I)
660 NEXT I
670 SE=SA/SF
680 VJ=V/(N-3)
690 CR=SQR(SE)
700 CC=SQR(VJ*ST/(SL*ST-SP^2))
710 CSC=SQR(VJ*SL/(SL*ST-SP^2))
720 SR=VJ/N+MX^2*CC+(SXSQ/N)^2*CSC^2-2*MX*(SXSQ/N)*SP*VJ/(SL*ST-SP^2)
730 F=(SK-VJ*(N-2))/(2*VJ)
740 PRINT
750 PRINT"Summary"
760 PRINT
770 PRINT"The intercept is "TAB(60) INT(S*1000+.5)/1000
780 PRINT "The coefficient of X is "TAB(60) INT(P*1000+.5)/1000
790 PRINT "The coefficient of X squared is "TAB(60) INT(M*1000+.5)/1000
800 PRINT "The variance is "TAB(60) INT(V*1000+.5)/1000
810 PRINT "The confidence limit of the estimate is "TAB(60) INT(SQR(VJ)*1000+.5)
/1000
820 PRINT "The confidence limit of the intercept is "TAB(60) INT (SQR(SR)*1000+.
5)/1000
830 PRINT "The confidence limit of the coefficient of X is "TAB(60) INT(CC*1000+
.5)/1000
840 PRINT "The confidence limit of the coefficient of X squared is " TAB(60) INT
(CSC*1000+.5)/1000
850 PRINT "The correlation coefficient is "TAB(60) INT(CR*1000+.5)/1000
860 PRINT "The F value is " TAB(60)INT(F*1000+.5)/1000
870 PRINT
880 PRINT TAB(10)"X";TAB(20)"Yobs";TAB(30)"Ycalc"
890 FOR I=1 TO N
900 PRINT TAB(10)X(I);TAB(20) Y(I);TAB(30)INT(Z(I)*1000+.5)/1000
910 NEXT I
920 PRINT
930 END
```

MINITAB command

REGRESS C3 on **2** predictors **C1** and **C2**. (This assumes that the values of *X* are in column 1, *X*-squared in column 2 and *Y* in column 3.)

This command gives the regression equation, the coefficients of *X* and *X*-squared, their standard deviations and *t*-ratios, the correlation coefficient and an analysis of variance.

1.4 THREE-VARIABLE REGRESSION

This program fits data to an equation of the form

$$Z=A+BX+CY$$

It calculates the values of the coefficients, *A, B*, and *C*, their confidence limits, the confidence limit of the estimate, and the correlation coefficient. It finally presents a table of *X, Y* the observed values of *Z* and the values of *Z* calculated using the regression equation. It can be used in Chapters 5 and 7.

Operation (using tablet disintegration data from Table 7.1).

Display	Response
Three-variable regression $Z=A+BX+CY$ Type in title	
	Tablet disintegration
Type in number of sets of data, and press RETURN	4
Type in data in order X1, Y1, Z1, X2, Y2, . . ., pressing RETURN after each entry	
	100 2.5 500 300 2.5 1070 100 7.5 140 300 7.5 640

Number	*X*	*Y*	*Z*
1	100	2.5	500
2	300	2.5	1070
3	100	7.5	140

4	300	7.5	640

Which set needs changing?
Type 0 if all correct.

0

Summary

Intercept is	447.5
Coefficient of X is	2.675
Coefficient of Y is	−79
Variance is	122.5
Confidence limit of the estimate is	3.5
Confidence limit of the intercept is	2756.25
Confidence limit of the X coefficient is	0.175
Correlation coefficient is	0.931
Confidence limit of the Y coefficient is	7
F value is	180.01

X	Y	Z_{obs}	Z_{calc}
100	2.5	500	517.5
300	2.5	1070	1052.5
100	7.5	140	122.5
300	7.5	640	657.5

```
100 REM three variable regression
110 PRINT "Three.variable regression"
120 PRINT
130 PRINT"Z = A + BX + CY"
140 PRINT
150 PRINT"Type in title"
160 INPUT T$
170 PRINT
180 PRINT"Type in number of sets of data, and press RETURN"
190 INPUT N
200 PRINT
210 DIM E(N):DIM G(N): DIM V(N): DIM W(N): DIM X(N):DIM Y(N): DIM Z(N)
220 PRINT "Type in data in order X1, Y1, Z1, X2, Y2..., pressing RETURN after ea
ch entry
230 PRINT
240 FOR J=1 TO N
250 INPUT X(J)
260 INPUT Y(J)
270 INPUT Z(J)
280 NEXT J
290 PRINT "NUMBER";TAB(20)"X";TAB(30)"Y";TAB(40)"Z"
300 FOR J=1 TO N
310 PRINT J;TAB(20) X(J); TAB(30)Y(J); TAB(40)Z(J)
320 NEXT J
330 PRINT
340 PRINT "Which set needs changing? Type 0 if all correct"
350 INPUT CH:IF CH=0 THEN 410
360 PRINT "New values of "CH
370 INPUT X(CH)
380 INPUT Y(CH)
390 INPUT Z(CH)
```

```
400 GOTO 330
410 FOR J=1 TO N
420 SX=SX+X(J):REM sx=sum of x
430 SXSQ=SXSQ+(X(J)*X(J)):REM sxsq=sum of x squared
440 SY=SY+Y(J):REM sy=sum of y
450 SYSQ=SYSQ+(Y(J)*Y(J)):REM sysq=sum of y squared
460 SZ=SZ+Z(J):REM sz=sum of z
470 SZSQ=SZSQ+(Z(J)*Z(J)):REM szsq=sum of z squared
480 XYSUM=XYSUM+(X(J)*Y(J)):REM xysum=sum of products of x and y
490 XZSUM=XZSUM+(X(J)*Z(J)):REM xzsum=sum of products of x and z
500 YZSUM=YZSUM+(Y(J)*Z(J)):REM yzsum=sum of products of y and z
510 NEXT J
520 H=(N*SXSQ-SX*SX)*(N*YZSUM-SY*SZ)
530 K=(N*XYSUM-SX*SY)*(N*XZSUM-SX*SZ)
540 L=(N*SXSQ-SX*SX)*(N*SYSQ-SY*SY)-(N*XYSUM-SX*SY)^2
550 M=(H-K)/L :REM m=coefficient of y
560 P=((N*XZSUM-SX*SZ)-M*(N*XYSUM-SX*SY))/(N*SXSQ-SX*SX):REM p=coefficient of x
570 S=(SZ-M*SY-P*SX)/N:REM s=intercept
580 PRINT
590 MX=SX/N:REM mx=mean value of x
600 MY=SY/N:REM my=mean value of y
610 MZ=SZ/N:REM mz=mean value of z
620 PRINT
630 FOR J=1 TO N
640 REM w is calculated value of z, using coefficents from equation
650 W(J)= S+P*X(J)+M*Y(J)
660 V(J)=(W(J)-Z(J))^2
670 SV=SV+V(J)
680 E(J)=(W(J)-MZ)^2
690 SE=SE+E(J)
700 G(J)=(Z(J)-MZ)^2
710 SG=SG+G(J)
720 NEXT J
730 VJ=SV/(N-3)
740 VE=SE/SG
750 PRINT
760 CR=SQR(VE):REM cr=correlation coefficient
770 PRINT
780 SK=SZSQ-SZ*SZ/N
790 SL=SXSQ-SX*SX/N
800 ST=SYSQ-SY*SY/N
810 SM=XZSUM-SX*SZ/N
820 SN=YZSUM-SY*SZ/N
830 SP=XYSUM-SX*SY/N
840 CC=VJ*ST/(SL*ST-SP^2)
850 CSC=VJ*SL/(SL*ST-SP^2)
860 SR=VJ/N+MX^2*CC+MY^2*CSC-((2*MX*MY*SP*VJ)/(SL*ST-SP*SP))
870 F=(SK-VJ*(N-2))/(2*VJ):REM f=f value
880 PRINT "Summary"
890  PRINT
900 PRINT "Intercept = "TAB(50) INT(S*1000+.5)/1000
910 PRINT "Coefficient of X is"TAB(50) INT(P*1000+.5)/1000
920 PRINT "Coefficient of Y is "TAB(50) INT(M*1000+.5)/1000
930  PRINT"Variance = "TAB(50) INT(VJ*1000+.5)/1000
940 PRINT "Confidence limit of the estimate is "TAB(50) INT (SQR(VJ)*1000+.5)/10
00
950  PRINT "Confidence limit of the intercept is "TAB(50) INT(SR*1000+.5)/1000
960  PRINT"Confidence limit of the X coefficient is "TAB(50) INT(SQR(CC)*1000+.5
)/1000
970 PRINT "Correlation coefficient is "TAB(50) INT(CC*1000+.5)/1000
980 PRINT "Confidence limit of the y coefficient is "TAB(50) INT(SQR(CSC)*1000+.
5)/1000
990  PRINT "F value is "TAB(50) INT(F*1000+.5)/1000
1000 PRINT
1010 PRINT TAB(10)"X";TAB(20)"Y";TAB(30)"Z obs";TAB(40)"Z calc"
1020 PRINT
1030 FOR J=1 TO N
1040 PRINT TAB(10)X(J);TAB(20)Y(J);TAB(30)Z(J);TAB(40)W(J)
1050 NEXT J
1060 END
```

MINITAB command

REGRESS C3 on **2** predictors **C1** and **C2**. (This assumes that values of X are in column 1, Y in column 2 and Z in column 3.)

This command gives the regression equation, the coefficients of X and Y, their standard deviations and t-ratios, the correlation coefficient and an analysis of variance.

1.5 THE DETERMINANT OF A (3×3) MATRIX

This program may be found useful in Chapter 6.

Operation, using data from Table 6.9.

Display	Response
Determinant of a (3×3) matrix Put in each element in turn, pressing RETURN after each entry.	
Row 1	
1	−0.8557
2	0.8949
3	−1.1152
Row 2	
1	1.0982
2	−1.0794
3	0.2984
Row 3	
1	−0.2402
2	0.1844
3	0.8168
−0.856 0.895 −1.115 1.098 −1.079 0.298 −0.240 0.184 0.817	
Determinant=0.002057	

```
100 PRINT "Determinant of a (3x3) matrix"
110 DIM B(3,3)
120 REM input of data
130 PRINT"Put in each element in turn, pressing RETURN after each entry"
140 FOR I=1 TO 3
150 PRINT "Row "I
160 FOR J=1 TO 3
170 PRINT J
180 INPUT B(I,J)
190 NEXT J
200 NEXT I
```

```
210 PRINT
220 PRINT "The matrix is:-"
230 PRINT
240 PRINT TAB(10)B(1,1);TAB(20)B(1,2);TAB(30)B(1,3)
250 PRINT TAB(10)B(2,1);TAB(20)B(2,2);TAB(30)B(2,3)
260 PRINT TAB(10)B(3,1);TAB(20)B(3,2);TAB(30)B(3,3)
270 PRINT
280 REM calculation of determinant by Cramer's rule
290 SUM=B(1,1)*(B(2,2)*B(3,3)-B(3,2)*B(2,3))
300 SUM=SUM-B(1,2)*(B(2,1)*B(3,3)-B(3,1)*B(2,3))
310 SUM=SUM+B(1,3)*(B(2,1)*B(3,2)-B(3,1)*B(2,2))
320 PRINT "Determinant = "INT(SUM*1000+.5)/1000
330 END
```

Associated MINITAB commands

MINITAB can add, subtract, multiply, transpose and invert matrices. The command:

EIGEN for **M1** put values in **C2** put vectors in **M2**

calculates eigenvalues and eigenvectors for a symmetrical matrix. The eigenvalues are stored in column 2 and the eigenvectors as columns of matrix M2.

1.6 THE DETERMINANT OF A (4×4) MATRIX

This program may be useful in Chapter 6.

Operation, using data from Table 6.10.

Display	Response
Determinant of a (4×4) matrix Put in each element in turn, pressing RETURN after each entry	
Row 1	
1	1.000
2	0.995
3	−0.944
4	−0.901
Row 2	
1	0.995
2	1.000
3	−0.946
4	0.882
Row 3	
1	−0.944
2	−0.946
3	1.000

```
4                                  0.800


Row 4
1                                 -0.901
2                                 -0.882
3                                  0.800
4                                  1.000

  1.000    0.995   -0.944   -0.901
  0.995    1.000   -0.946   -0.882
 -0.944   -0.946    1.000    0.800
 -0.901   -0.882    0.800    1.000

Determinant=0.0038
```

```
100 PRINT "Determinant of a (4x4) matrix"
110 DIM B(4,4)
120 REM input of data
130 PRINT"Put in each element in turn, pressing RETURN after each entry"
140 FOR I=1 TO 4
150 PRINT "Row "I
160 FOR J=1 TO 4
170 PRINT J
180 INPUT B(I,J)
190 NEXT J
200 NEXT I
210 PRINT
220 PRINT "The matrix is:-"
230 PRINT TAB(10)B(1,1);TAB(20)B(1,2);TAB(30)B(1,3);TAB(40)B(1,4)
240 PRINT TAB(10)B(2,1);TAB(20)B(2,2);TAB(30)B(2,3);TAB(40)B(2,4)
250 PRINT TAB(10)B(3,1);TAB(20)B(3,2);TAB(30)B(3,3);TAB(40)B(3,4)
260 PRINT TAB(10)B(4,1);TAB(20)B(4,2);TAB(30)B(4,3);TAB(40)B(4,4)
270 PRINT
280 REM calculation of determinant by Cramer's rule
290 SUM = B(1,1)*(B(2,2)*(B(3,3)*B(4,4)-B(4,3)*B(3,4))-B(2,3)*(B(3,2)*B(4,4)-B(4
,2)*B(3,4))+B(2,4)*(B(3,2)*B(4,3)-B(4,2)*B(3,3)))
300 SUM = SUM - B(1,2)*(B(2,1)*(B(3,3)*B(4,4)-B(4,3)*B(3,4))-B(2,3)*(B(3,1)*B(4,
4)-B(4,1)*B(3,4))+B(2,4)*(B(3,1)*B(4,3)*-B(4,1)*B(3,3)))
310 SUM = SUM + B(1,3)*(B(2,1)*(B(3,2)*B(4,4)-B(4,2)*B(3,4))-B(2,2)*(B(3,1)*B(4,
4)-B(4,1)*B(3,4))+B(2,4)*(B(3,1)*B(4,2)-B(4,1)*B(3,2)))
320 SUM = SUM - B(1,4)*(B(2,1)*(B(3,2)*B(4,3)-B(4,2)*B(3,3))-B(2,2)*(B(3,1)*B(4,
3)-B(4,1)*B(3,3))+B(2,3)*(B(3,1)*B(4,2)-B(4,1)*B(3,2)))
330 PRINT "Determinant = "INT(SUM*1000+.5)/1000
340 PRINT
350 END
```

For associated MINITAB commands, see 1.5.

1.7 THREE-FACTOR, TWO-LEVEL FACTORIAL DESIGN

This program calculates main effects, interactions and Yates' treatment, using data derived from a three-factor, two-level factorial design. Note that the data from the experiments must be arranged in standard order.

This program is useful in Chapter 4.

Operation (using data from Table 4.6).

Display	Response
Three-factor, two-level design	
Experiments must be in standard order.	
Enter response for each experiment when prompted and press RETURN	
Experiment (1)	
	1.6
Experiment a	
	5.3
Experiment b	
	3.4
Experiment ab	
	6.6
Experiment c	
	2.6
Experiment ac	
	3.6
Experiment bc	
	3.0
Experiment abc	
	7.0

Experiment	Response	Column 1	Column 2	Column 3	Effect	Mean square
(1)	1.6	6.9	16.9	33.1	—	—
a	5.3	10.0	16.2	11.9	2.975	17.7
b	3.4	6.2	6.9	6.9	1.725	5.95
ab	6.6	10.0	5.0	2.5	0.625	0.78
c	2.6	3.7	3.1	−0.70	−0.175	0.061
ac	3.6	3.2	3.8	−1.9	−0.475	0.45
bc	3.0	1.0	−0.5	0.7	0.175	0.06
abc	7.0	4.0	3.0	3.5	0.875	1.53

Another set of responses? (y/n)

```
100 REM three factor, two level factorial design
110 REM Calculation of main effects, interactions, Yate's treatment for ANOVA
120 CLS
130 PRINT "Three factor, two level design"
140 PRINT
150 PRINT
160 PRINT "Experiments must be in standard order"
170 PRINT
180 PRINT "Enter response for each experiment when prompted and press RETURN
190 PRINT "Experiment (1)"
200 INPUT N1
210 PRINT "Experiment a"
220 INPUT N2
230 PRINT "Experiment b"
```

```
240 INPUT N3
250 PRINT "Experiment ab"
260 INPUT N4
270 PRINT "Experiment c"
280 INPUT N5
290 PRINT "Experiment ac"
300 INPUT N6
310 PRINT "Experiment bc"
320 INPUT N7
330 PRINT "Experiment abc"
340 INPUT N8
350 REM designate numbers in column 1 as z1,z2 etc
360 Z1=N1+N2: Z2=N3+N4: Z3=N5+N6:Z4=N7+N8:Z5=N2-N1:Z6=N4-N3:Z7=N6-N5:Z8=N8-N7
370 REM designate column 2 y1,y2 etc
380 Y1=Z1+Z2:Y2=Z3+Z4:Y3=Z5+Z6:Y4=Z7+Z8:Y5=Z2-Z1:Y6=Z4-Z3:Y7=Z6-Z5:Y8=Z8-Z7
390 REM designate column 3 as x1, x2 etc
400 X1=Y1+Y2:X2=Y3+Y4:X3=Y5+Y6:X4=Y7+Y8:X5=Y2-Y1:X6=Y4-Y3:X7=Y6-Y5:X8=Y8-Y7
410 REM main efffects, interactions in column 4 (e1,e2 etc)
420 REM mean squares in column 5 (m1,m2 etc)
430 E1=X1/4:M1=(X1)^2/8
440 E2=X2/4:M2=(X2)^2/8
450 E3=X3/4:M3=(X3)^2/8
460 E4=X4/4:M4=(X4)^2/8
470 E5=X5/4:M5=(X5)^2/8
480 E6=X6/4:M6=(X6)^2/8
490 E7=X7/4:M7=(X7)^2/8
500 E8=X8/4:M8=(X8)^2/8
510 PRINT
520 PRINT "Experiment";TAB(12)"Response";TAB(22)"Column";TAB(32)"Column";TAB(42)
"Column";TAB(52)"Effect";TAB(62)"Mean"
530 PRINT TAB(25)"1";TAB(35)"2";TAB(45)"3";TAB(61)"square"
540 PRINT
550 PRINT TAB(6)"(1)";TAB(14)N1;TAB(24)Z1;TAB(34)Y1;TAB(44)X1;TAB(55)"-";TAB(65)
"-"
560 PRINT TAB(7)"a";TAB(14)N2;TAB(24)Z2;TAB(34)Y2;TAB(44)X2;TAB(54)E2;TAB(64)M2
570 PRINT TAB(7)"b";TAB(14)N3;TAB(24)Z3;TAB(34)Y3;TAB(44)X3;TAB(54)E3;TAB(64)M3
580 PRINT TAB(6)"ab";TAB(14)N4;TAB(24)Z4;TAB(34)Y4;TAB(44)X4;TAB(54)E4;TAB(64)M4
590 PRINT TAB(7)"c";TAB(14)N5;TAB(24)Z5;TAB(34)Y5;TAB(44)X5;TAB(54)E5;TAB(64)M5
600 PRINT TAB(6)"ac";TAB(14)N6;TAB(24)Z6;TAB(34)Y6;TAB(44)X6;TAB(54)E6;TAB(64)M6
610 PRINT TAB(6)"bc";TAB(14)N7;TAB(24)Z7;TAB(34)Y7;TAB(44)X7;TAB(54)E7;TAB(64)M7
620 PRINT TAB(5)"abc";TAB(14)N8;TAB(24)Z8;TAB(34)Y8;TAB(44)X8;TAB(54)E8;TAB(64)M
8
630 LPRINT"Experiment";TAB(12)"Response";TAB(22)"Column";TAB(32)"Column";TAB(42)
"Column";TAB(55)"Effect";TAB(65)"Mean"
640 LPRINT TAB(25)"1";TAB(35)"2";TAB(45)"3";TAB(64)"square"
650 LPRINT
660 LPRINT TAB(6)"(1)";TAB(14)N1;TAB(24)Z1;TAB(34)Y1;TAB(44)X1;TAB(55)"-";TAB(65
)"-"
670 LPRINT TAB(7)"a";TAB(14)N2;TAB(24)Z2;TAB(34)Y2;TAB(44)X2;TAB(54)E2;TAB(64)M2
680 LPRINT TAB(7)"b";TAB(14)N3;TAB(24)Z3;TAB(34)Y3;TAB(44)X3;TAB(54)E3;TAB(64)M3
690 LPRINT TAB(6)"ab";TAB(14)N4;TAB(24)Z4;TAB(34)Y4;TAB(44)X4;TAB(54)E4;TAB(64)M
4
700 LPRINT TAB(7)"c";TAB(14)N5;TAB(24)Z5;TAB(34)Y5;TAB(44)X5;TAB(54)E5;TAB(64)M5
710 LPRINT TAB(6)"ac";TAB(14)N6;TAB(24)Z6;TAB(34)Y6;TAB(44)X6;TAB(54)E6;TAB(64)M
6
720 LPRINT TAB(6)"bc";TAB(14)N7;TAB(24)Z7;TAB(34)Y7;TAB(44)X7;TAB(54)E7;TAB(64)M
7
730 LPRINT TAB(5)"abc";TAB(14)N8;TAB(24)Z8;TAB(34)Y8;TAB(44)X8;TAB(54)E8;TAB(64)
M8
740 LPRINT:LPRINT
750 PRINT
760 PRINT "Another set of responses?  (y/n)"
770 INPUT Q1$
780 IF Q1$="y" THEN 100
790 STOP
```

Appendix 2: Mathematical tables

Full sets of mathematical tables for techniques used in this book are widely available in reference sources and textbooks on statistics. Hence the only tables given here are those which are referred to in the worked examples in the text.

2.1 CUMULATIVE NORMAL DISTRIBUTION

Cumulative area under the normal distribution curve (less or equal to z).

z	Area
1.65	0.90
1.96	0.95
2.58	0.99

2.2 t-DISTRIBUTION

Degrees of freedom	Two-sided One-sided	5% 2.5%	1% 0.5%
1		12.71	63.66
2		4.30	9.92
3		3.18	5.84
4		2.28	4.60
5		2.57	4.03
6		2.45	3.71
7		2.36	3.50
8		2.31	3.36
9		2.26	3.25

10	2.23	3.17
11	2.20	3.11
12	2.18	3.06
13	2.16	3.01
14	2.14	2.98
15	2.13	2.95
16	2.12	2.92
17	2.11	2.90
18	2.10	2.88
19	2.09	2.86
20	2.09	2.85

2.3 UPPER 5% VALUES OF THE *F* DISTRIBUTION

Degrees of freedom in denominator	Degrees of freedom in numerator				
	1	2	3	4	5
1	161	200	216	225	2301
2	18.5	19.0	19.2	19.2	19.3
3	10.1	9.55	9.28	9.12	9.01
4	7.71	6.94	6.59	6.39	6.26
5	6.61	5.79	5.41	5.19	5.05
6	5.99	5.14	4.76	4.53	4.39
7	5.59	4.74	4.35	4.12	3.97
8	5.32	4.46	4.07	3.84	3.69
9	5.12	4.26	3.86	3.63	3.48
10	4.96	4.10	3.71	3.48	3.33
15	4.54	3.68	3.29	3.06	2.90
20	4.35	3.49	3.10	2.87	2.71
25	4.24	3.39	2.99	2.76	2.60
30	4.17	3.32	2.92	2.69	2.53

2.4 UPPER 1% VALUES OF THE *F* DISTRIBUTION

Degrees of freedom in denominator	Degrees of freedom in numerator				
	1	2	3	4	5
1	4052	4999	5403	5625	5764
2	98.5	99.0	99.17	99.25	99.30
3	34.12	30.82	29.46	28.71	28.24
4	21.20	18.00	16.69	15.98	15.52
5	16.26	13.27	12.06	11.39	10.97

6	13.75	10.92	9.78	9.15	8.75
7	12.25	9.55	8.45	7.85	7.46
8	11.26	8.65	7.59	7.01	6.63
9	10.56	8.02	6.99	6.42	6.06
10	10.04	7.56	6.55	5.99	5.64
15	8.68	6.36	5.42	4.89	4.56
20	8.10	5.85	4.94	4.43	4.10
25	7.77	5.57	4.68	4.18	3.85
30	7.56	5.39	4.51	4.02	3.70

Index